AUTOIMMUNITY AND THE EYE

This book is dedicated to the memory of
Cora Verhagen of Amsterdam,
The Netherlands
and
Ortwin Rott of Paris, France.

Autoimmunity and the Eye

EDITORS

MANFRED ZIERHUT
University Eye Clinic
Tübingen

MICHAEL RAIZMAN
New England Eye Center
Tufts University School of Medicine
Boston, MA

HANS-JÜRGEN THIEL
University Eye Clinic
Tübingen

ÆOLUS PRESS

SCIENCE PUBLISHERS

BUREN

THE NETHERLANDS

ISBN 90-7043-014-2
Æolus Press Science Publishers, P.O. Box 740, 4116 ZJ Buren, The Netherlands

Orders
Swets & Zeitlinger Publishers, P.O. Box 825, 2160 SZ Lisse, The Netherlands
USA and Canada
Swets & Zeitlinger Publishers, P.O. Box 613, Royersford, PA 19468, USA

Æolus Press is a member of Swets & Zeitlinger Publishers

Typeset by Perfect Service, Schoonhoven and printed in The Netherlands by Krips BV, Meppel on acid-free paper.

Contents

T. E. W. FELTKAMP

Autoimmunity and autoimmune diseases of the eye

Introduction

Although Uhlenhuth showed in 1903[1] that rabbits, immunized with bovine lens proteins, started to produce antibodies to antigens which also existed in the rabbit eye, this description of an autoimmune phenomenon drew less attention than the concept of "horror autotoxicus" which was introduced by Ehrlich and Morgenroth two years earlier, based on the phenomenon that sheep do not make antibodies against their own blood cells.[2]

The present article will introduce the concepts tolerance, autoimmunity and autoimmune disease without discussing experimental details. These will be described in subsequent papers in this book. It will be underlined that immunologic defence against self is a normal situation. A small quantitative and/or qualitative increase of this defence may lead to autoimmune disease. The definition of autoimmune diseases is therefore based on practical notions.

Autoimmune diseases can be divided into several groups. This also holds for the ocular autoimmune diseases. It will be shown that most of the autoimmune diseases of the eye represent a facet of a generalized disease and that the pathogenetic mechanisms of most of the diseases confined to the eye are poorly understood. Several exciting experimental autoimmune diseases of the eye can be induced. These experimental autoimmune diseases, however, only partially reflect the spontaneous human ocular diseases.

Tolerance

Tolerance is looked upon as a specific immunological non-reactivity. This means that we are dealing with non-reactivity against some antigenic peptides in the presence of normal reactivity against other antigens.

Since the observation by Owen in 1945,[3] who showed that dizygotic cattle siblings, having shared a common placental circulation, could freely exchange skin transplants later in life, immunologists became aware of the fact that tolerance has something to do with early recognition of antigens. The adult cows and bulls might be considered a culture medium for stem cells of their genetically different siblings. This situation is known as chimerism. These observations were reproduced in an experimental situation by Billingham et al. in 1953.[4]

Von Boehmer[5] extended these observations while studying mice which had received the genetic information for receptors on the surface of cytotoxic T cells (CTL) recognizing antigenic peptides. Making transgenic mice with T cell receptors (TCR) which react with a male mouse protein called HY, he observed that the female offspring developed a normal thymus and cytotoxic T cells (CTL) against HY. The male mice, however, showed an underdeveloped thymus and no anti-HY CTL's. The clonal deletion in the males affected mainly the $CD4^+/CD8^+$ $T\alpha\beta$ cells which reacted with a high affinity with the abundant HY antigens in the male thymus.

It is now generally accepted that when many equally antigenic peptides are bound with a high affinity to TCR's early in life, this mostly causes the death of this recognizing T cell clone. If such a confrontation between antigens and the immune system persists, tolerance for such antigens will be obtained. Such a process is present in the autologous situation: an excess of self antigens permanently confronts newly developing cells of the immune system. This leads to the deletion of anti-self T cell clones.

For B lymphocyte tolerance the situation is comparable, although these cells are often not killed by an early, intense and continuous recognition of antigens but become "anergic", since they are becoming unreactive to stimulation.[6]

It is of utmost importance to realize that it is far easier to induce tolerant T cells than tolerant B cells.[7] Such T cell tolerance also persists much longer than tolerance of B cells. As a consequence, human beings are full of B cells which are capable of reacting with certain autologous peptides, only they do not since they lack the help of proper T helper cells. From the moment that such T helper cells appear, the anti-self B cells become activated. This results in autoimmune phenomena.

Autoimmunity

Not all antigenic peptides are properly presented with high affinity to TCR, not all peptides occur in a high enough concentration and not all, e.g. sperm antigens, are present at the time of development of the immune system. It is therefore not astonishing that many anti-self lymphocytes are normally present.

It should also be kept in mind that most antigens consist of many different antigenic determinants or epitopes. Immunologic defence against a foreign red cell or a bacterium deals, in fact, with many distinct antibodies against separate epitopes. It is obvious that the overall final reactivity against such an antigen is the resultant of many individual reactivities and tolerances against these distinct epitopes.

In conclusion, it can be stated that the balance between tolerance and autoimmunity is delicate and that the border between tolerance and autoimmunity is vague. There are therefore many degrees of autoimmunity.

Everyone is *autoimmunopotent*. This means that everyone has not only autoreactive B cells waiting for the help of proper T cells, but also produces many anti-self antibodies. A glance at a Western blot, showing our own serum full of distinct antibodies reacting with our own cytoplasmic and nuclear antigens, is revealing. Serologists working in routine laboratories were already aware of this phenomenon for a long time. They decided not to report the presence of autoantibodies, like rheumatoid factors, unless the titer is above the level which is present in 95% of the healthy population. The presence of autoantibodies is therefore completely normal, unless they occur in high titers such as are found in only 5% of mankind.

Many of us are *autoimmunoresponsive*. One in four of the women over 60 years of age produces autoantibodies in high titers against antigens in thyroid or gastric mucosa. Most of these women show no clinical abnormalities and their general practitioner would consider them completely healthy. If, however, a morbid anatomist would have been able to view their thyroid or gastric tissue, the existence of lymphocellular infiltrations would have been noted. Subclinical autoimmune phenomena are thus frequent.

Three percent of the population has a real *autoimmune disease.* Autoimmune diseases are therefore not so rare at all. Among these autoimmune diseases, rheumatoid arthritis is the most frequent. Another large group represents the diseases of the thyroid. Practically all thyroid diseases with hyper- or hypofunction, with or without nodules, are of autoimmune origin. The third large group is that of the hepatic autoimmune diseases.

Autoimmune disease

Because the border between tolerance and autoimmunity is vague, the definition of autoimmune disease should be clear and practical. In my personal opinion the best definition is: *A disease characterized by the presence of autoantibodies in a significantly increased frequency and with significantly increased titers if compared with healthy controls.*

As with most definitions, this definition is certainly not perfect: Syphilis with its antibodies to cardiolipins, which are in fact autoantibodies, will generally not be considered to be an autoimmune disease. The definition does not mention cellular autoimmunity, although in experimental models of autoimmune diseases of the nervous system (EAE), the thyroid (EAT), the uvea (EAU) and other organs, the disease is only transferable with T cells. Limiting the definition to humoral autoimmunity is done for practical reasons. To my knowledge, there are no human autoimmune diseases in which autoantibodies in increased frequencies and titers are absent. Furthermore, it is evident that antibody determination is far easier than the demonstration of cellular autoimmune reactions.

Diseases with inflammatory aspects of unknown origin in which autoantibodies are not present in frequencies or titers which are significantly increased, should not be considered as autoimmune diseases just because they resemble experimental autoimmune diseases. Multiple sclerosis, uveitis of unknown origin or ankylosing spondylitis can therefore not yet be classified as autoimmune diseases.

It should be kept in mind that most autoantibodies do not cause disease phenomena. Only antibodies against red cells, leukocytes, platelets, acetylcholine receptors, receptors for thyroid stimulating hormone, intrinsic factor and glomerular basement membrane are clearly known to be harmful.

Primary versus secondary autoimmune diseases

Autoimmune diseases can be divided into diseases with an impaired immune system and diseases in which an autoantigen is somewhat altered but the immune apparatus is completely normal. The diseases with normal antigens are called primary, those with a normal immune system secondary (see Fig 1). In Table 1, the differences between the primary and secondary autoimmune diseases are listed.

The cause of the *primary autoimmune diseases* is completely unknown. The diseases are therefore also known as idiopathic autoimmune diseases. In these diseases, B cells able to produce autoantibodies receive help from T cells for unknown reasons. Women are in general somewhat more immunocompetent than men. The elderly are supposed to have a less controlled immune system than youngsters and adults. These unsatisfactory reasons are brought forward to explain the fact that primary autoimmune diseases are more frequently seen in elderly women than in young boys. The increased T cell help is directed to many different B cell clones. Thus, patients with one primary autoimmune disease often show

TABLE 1 Primary versus secondary autoimmune diseases (AID).

Primary autoimmune diseases	*Secondary autoimmune diseases*
Cause unknown	Cause known or suspected
Autoantigens normal	Autoantigens altered or Mimicry with exogenous antigen or Leakage of sequestered autoantigens
Excess of T cell help	Immune system normal
Permanent disease	Transient autoimmune phenomena
Can be treated	Curable autoimmune phenomena
Mainly women	Both sexes
Rarely in children	All ages
Clinical overlap with other primary AID	No overlap
Serological overlap with other primary AID	No overlap
Familial predisposition	Rarely

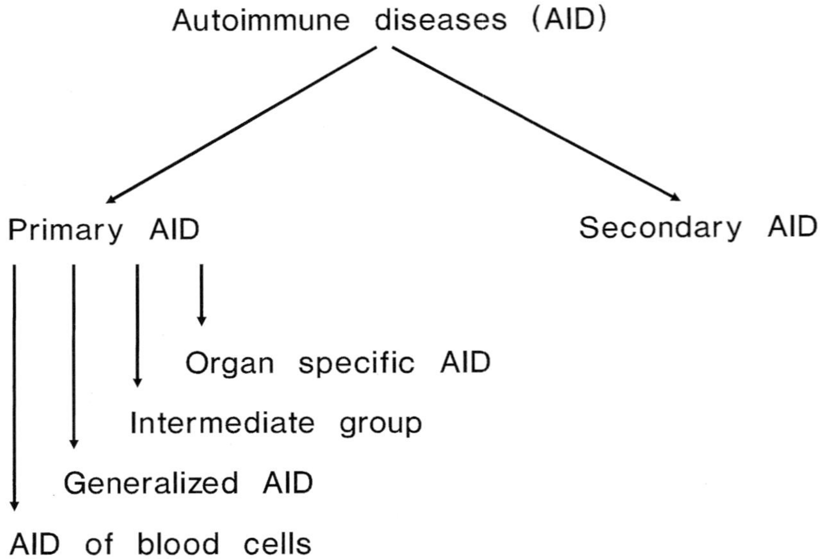

FIGURE 1 Subdivision of the autoimmune diseases.

symptoms of other primary autoimmune diseases at the same time. This also holds true for their relatives.

The *secondary autoimmune diseases* have their onset mostly after a certain event such as an operation, trauma, infection, infarction or drug intake. Such an event probably leads to the appearance of new antigenic determinants on autologous cells or tissues, resulting in autoreactivity by a normal immune system. It is also possible that previously well sequestered antigens suddenly reach the immune system. This might be the case for spermal antigens after vasectomy or for lens proteins after extracapsular lens extraction. In general, the secondary autoimmune diseases are transient. As soon as the altered antigen has disappeared, the autoimmune reaction is arrested. If tissue destruction was considerable during the period of autoimmunity, the resulting deficiency may not be transient even when the autoimmune inflammation has come to an end. An example of this is juvenile diabetes mellitus in which the β cells of the islets of Langerhans are destroyed by the autoimmune inflammation secondary to the altered antigens, probably due to a viral infection. The immune system of these patients is completely normal. The chance of developing another autoimmune disease is therefore not increased. These diseases have no prevalence for elderly women.

The secondary autoimmune diseases include juvenile insulin-dependent diabetes mellitus of type Ia, male sterility due to anti-spermatozoal antibodies, phacogenic uveitis, post-myocardial infarction syndrome, post-pericardiotomy syndrome, rheumatic carditis, Goodpasture's syndrome and Wegener's granulomatosis.

Organ-specific versus generalized autoimmune diseases

Around 1960 it became clear that the primary autoimmune diseases could be divided into organ-specific and generalized autoimmune diseases (see Fig. 1).

The organ-specific diseases are more or less confined to one organ (thyroid, gastric mucosa and adrenocortex). These diseases show an intense clinical and serological overlap.

The generalized diseases manifest their effects all over the body. Systemic lupus erythematosus (SLE) is the classical example of these diseases. Other diseases showing a clinical and serological overlap with SLE, like rheumatoid arthritis, scleroderma, Sjögren's syndrome, CREST syndrome and mixed connective tissue disease (MCTD), are also looked upon as generalized autoimmune diseases.

The autoimmune diseases of the blood cells: autoimmune hemolytic anemia, autoimmune granulocytopenia, lymphocytopenia and thrombocytopenia and pure red cell anemia, are mostly not considered generalized autoimmune diseases although they are often seen as symptoms of these diseases. They are grouped separately as autoimmune diseases of the blood cells.

If Sjögren's syndrome were not characterized by the presence of many distinct antinuclear and anticytoplasmic antibodies also occurring in SLE, it would possibly have been classified as belonging to the organ- specific autoimmune diseases. Although 40% of patients with myasthenia gravis or chronic active hepatitis also have antinuclear antibodies, these autoimmune diseases are not looked upon as generalized autoimmune diseases. Together with primary biliary cirrhosis, pemphigus and pemphigoid they are classified as an intermediate group.

Primary autoimmune diseases of the eye

These diseases are listed in Table 2. The only possibly primary, organ-specific autoimmune disease of the eye could be *Graves' oph-*

thalmopathy. This disease, which is a part of autoimmune hyperthyroiditis, is an inflammation in the eye muscles and the orbital connective tissue, leading to an increase of the orbital tissue mass and thereby to exophthalmos. The hyperthyroiditis is the result of autoantibodies to receptors for thyroid stimulating hormone (TSH). It is possible that the orbital inflammation is also caused by autoantibodies; however, neither the autoantigen nor the autoantibodies have been determined. The disease therefore only partially fulfils the definition of an autoimmune disease.

A generalized, primary autoimmune disease of the eye is *Sjögren's syndrome*, with autoantibodies directed to the SS-B or La antigen and the SS-A or Ro antigen. Other antinuclear antibodies and antibodies to the Sm antigen are also frequently observed in this syndrome.

Other generalized primary autoimmune diseases of the eye are: *SLE* with a vascular retinitis characterized by antinuclear antibodies especially directed to DNA, *rheumatoid arthritis* with a scleritis characterized by rheumatoid factors, being autoantibodies to the Fc part of IgG molecules, and by antiperinuclear factors, being antibodies to keratohyaline bodies in squamous mucous epithelium, and *juvenile chronic arthritis with a pauciarticular onset* with often severe bilateral uveitis, in which autoantibodies to a 15 kD nuclear antigen often occur.[8]

Primary autoimmune diseases of the eye of the group intermediate between organ-specific and generalized are *myasthenia gravis*, leading to ptosis and strabismus due to autoantibodies to acetylcho-

TABLE 2 Autoimmune diseases of the eye.

Primary autoimmune diseases	*Secondary autoimmune diseases*
Graves' ophthalmopathy	juvenile insulin-dependent diabetes mellitus
Sjögren's syndrome	Wegener's granulomatosis
SLE	Mooren's ulcer
rheumatoid arthritis	phacogenic uveitis
juvenile chronic arthritis	Vogt-Koyanagi-Harada disease
myasthenia gravis	sympathetic ophthalmia
pemphigus	
pemphigoid	

line receptors which promote invagination and thereby diminution of these receptors, *pemphigus* which can lead to keratitis due to autoantibodies to an intra-epithelial cellular kit substance and *pemphigoid*, a comparable disease with autoantibodies to the basal membrane of skin and cornea.

Secondary autoimmune diseases of the eye

In these diseases, listed in Table 2, a certain event took place before the onset of the disease. In *juvenile insulin-dependent diabetes mellitus* of type Ia, a viral infection probably affected the islets of Langerhans, leading to altered antigens in these islets which induce the production of autoantibodies. Although autoantibodies against all cells of the islets are produced, it is only the deficiency of the insulin-producing β cells which leads to clinical phenomena. The eye involvement consists of diabetic vasculopathy as a complication of this disease.

Another generalized autoimmune disease is *Wegener's granulomatosis*. This disease is only classified as a secondary autoimmune disease because the disease is often associated with, or preceded by, infections of the respiratory tract. It should be admitted, however, that the causal agent of the granulomatosis is not known. The disease is characterized by the presence of autoantibodies to proteinase 3 occurring in phagosomes of granulocytes.[9]

In *Mooren's ulcer*, which is probably caused by exogenous trauma or infection, autoantibodies against several corneal antigens are observed. One of the antigens possibly involved is a 54 kD protein recently identified as an aldehyde dehydrogenase.[10,11]

Phacogenic uveitis due to autoantibodies against lens proteins, evoked by the release of such proteins, was already mentioned above.

In *Vogt-Koyanagi-Harada disease*, possibly the result of a viral infection, autoantibodies against melanocytes are supposed to be present. However, the reports about these antibodies lack firm confirmation.

The last disease to be mentioned in this list is *sympathetic ophthalmia*. This syndrome has already been known for a long time. Nevertheless, strong evidence as to the pathogenesis, especially whether autoantibodies play a significant role, is lacking.

Experimental autoimmune diseases of the eye

The injection of retinal S-antigen (arrestin), interphotoreceptor retinoid-binding protein (IRBP), rhodopsin or recoverin together with Freund's adjuvant is able to induce experimental autoimmune posterior uveitis (EAU) in animals.[12] With melanin an experimental autoimmune anterior uveitis (EAAU) can be induced.[13]

These animals produce antibodies to these antigens although the diseases are mainly mediated by T lymphocytes. Often the experimental uveitis is accompanied by an inflammation of the pineal gland. In humans, however, these antibodies occur in the same frequencies and titers in uveitis patients as in healthy individuals. The same accounts for the cellular immunity to these antigens. As yet, no evidence has been reported for pineal gland involvement in human uveal disease.

The study of EAU has taught us a great deal about the pathogenesis of ocular inflammation. To consider these diseases as a model for human uveitis of unknown origin is, however, an oversimplification.[14]

Conclusion

In conclusion, the following statements can be made:

1. The borderline between autoimmunity in healthy individuals and autoimmune phenomena in patients with autoimmune diseases is vague.
2. The definition of autoimmune disease should therefore be practical.
3. Primary autoimmune diseases are not confined to the eye.
4. The few secondary autoimmune diseases which are confined to the eye are still poorly understood.
5. Experimental autoimmune diseases are different from human autoimmune diseases.

Summary

Tolerance against autoantigens is far from complete. The balance between tolerance and autoimmunity is delicate and the border between tolerance and autoimmunity is vague. Everyone is autoimmunopotent, many are autoimmunoresponsive and three percent of the population has an autoimmune disease.

An autoimmune disease is defined as a disease characterized by the presence of autoantibodies in a significantly increased frequency and with significantly increased titers if compared with healthy controls. The autoimmune diseases are divided into primary diseases, characterized by an increased help of T lymphocytes activating anti-self B lymphocytes, and secondary diseases in which the immune system is normal but some autoantigens are altered. The primary autoimmune diseases are divided into diseases of blood cells, organ-specific autoimmune diseases, generalized autoimmune diseases and an intermediate group between the latter two. None of the primary autoimmune diseases of the eye are confined to the eye itself. The few secondary autoimmune diseases which are confined to the eye are not well understood.

It should be kept in mind that the experimental ophthalmological autoimmune diseases are nice tools to study ophthalmological inflammations, but that the pathogenesis probably differs from human diseases of the eye.

References

1 Uhlenhuth P. Zur Lehre von der Unterscheidung verschiedener Eiweissarte mit Hilfe spezifischer Sera. In: Festschrift zum 60 Geburtstag von Robert Koch. Jena: Gustav Fischer, 1903:49-74.

2 Ehrlich P, Morgenroth J. II Ueber Hämolysine, Fünfte mitteilung. Berl Klin Wschr 1901;38:251.

3 Owen RD. Immunogenetic consequences of vascular anastomoses between bovine twins. Science 1945;102:400.

4 Billingham RE, Brent L, Medawar PB. "Actively acquired tolerance" of foreign cells. Nature 1953;172:603-6.

5 Von Boehmer H, Teh HS, Kisielow P. The thymus selects the useful, neglects the useless and destroys the harmful. Immunology Today 1989;10:57-61.

6 Nossal GJ. Molecular and cellular aspects of immunologic tolerance. Eur J Biochem 1991;202:729-37.

7 Weigle WO. Immunological unresponsiveness. Adv Immunol 1973;16:61-122.

8 Neuteboom GHG, Hertzberger-ten Cate R, De Jong J, Van den Brink HG, Feltkamp TEW. Antibodies to a 15kD nuclear antigen in patients with juvenile chronic arthritis and uveitis. Invest Ophthalmol Vis Sci 1992;33:1657-60.

9 Kallenberg CGM, Cohen Tervaert JW, Van der Woude FJ, Goldschmeding R, Von dem Borne AEGK, Weening JJ. Autoimmunity to lysosomal enzymes: new clues to vasculitis and glomerulonephritis. Immunology Today 1991;12:61-5.

10 Kruit PJ, Van der Gaag R, Broersma L, Kijlstra A. Autoimmunity against

corneal antigens. I. Isolation of a soluble 54kD corneal antigen. Curr Eye Res 1986;5:313-20.

11 Verjans GMGM, Verhagen C, Hoekzema R, Kijlstra A. Bovine corneal protein 54 K (BCP 54) shows strong homology to rat tumor-associated aldehyde dehydrogenase (T-ALDH). Curr Eye Res 1990;9:1217-8.

12 Gery I, Chanaud NP III, Anglade E. Recoverin is highly uveitogenic in Lewis rats. Invest Ophthalmol Vis Sci 1994;35:3342-5.

13 Broekhuyse RM, Kuhlmann ED, Winkens HJ, Van Vugt AHM. Experimental autoimmune anterior uveitis (EAAU), a new form of experimental uveitis. I. Induction by a detergent-insoluble intrinsic protein fraction of the retinal pigment epithelium. Exp Eye Res 1991;52:465-74.

14 Kijlstra A, Hoekzema R, Van der Lelij A, Doekes G, Rothova A. Humoral and cellular immune reactions against retinal antigens in clinical disease. Curr Eye Res 1990;9:85-9.

JÖRG REIMANN, ANDREAS KUHRÖBER AND REINHOLD SCHIRMBECK

Alternative processing of autoantigen may trigger activation of MHC class I-restricted, self-reactive T cells

Introduction

Responses of peripheral, immunocompetent T cells to autologous protein antigens are assumed to be of critical importance in the pathogenesis of autoimmune diseases. This idea is supported by many experimental models in which different autoimmune pathologies are inducible by T cells. Only scant evidence is available for a role of self-reactive T cells in the pathogenesis of 'spontaneous' autoimmune disease in man.

The nature of T cell tolerance and the conditions under which it is broken are unresolved problems.[1] The questions addressing this problem can be grouped under four subheadings. *(i)* Which developmental pathway generates self-reactive T cells? *(ii)* Which regulatory controls prevent priming of peripheral, potentially autoaggressive T cell clones? *(iii)* Can a change in processing or presentation of an autologous protein render it immunogenic for T cells? *(iv)* Which features of the autoantigen are of critical importance for its immunogenicity for autologous T cells?

The priming of self-reactive T cell responses

Little is known about the differentiation pathway from which self-reactive T cells emerge into the peripheral, immunocompetent T cell pool. Development of self-reactive T cells may be common or may represent a rare, accidental event. Self-reactive T cells may develop in the thymus or in extrathymic pathways. They may be primed by an autoantigen or a 'foreign' (*e.g.* microbial) antigen; hence, they may be selected to recognize, either exclusively an epitope

Abbreviations: ***MHC****, major histocompatibility complex;* ***APC****, antigen-presenting cell;* ***TCR****, T cell receptor for antigen;* ***ER****, endoplasmic reticulum;* ***OVA****, ovalbumin;* ***CTL****, cytotoxic T lymphocyte;* ***CPL****, compartment for peptide loading;* ***VLP****, virus-like particle;* ***S****, surface (antigen);* ***HBV****, hepatitis B virus;* ***hsp****, heat shock proteins;* ***mAb****, monoclonal antibody* of an autoantigen, or the epitope of a 'foreign' protein antigen and display a variable degree of cross-reactivity with the epitope of an autoantigen. Hence, the developmental origin of peripheral, self-reactive T cells is still obscure.

Immunocompetent, self-reactive T cells may arise frequently but may be subjected to tight 'suppressive' regulatory controls. Functional activation of these T cells could result from 'excessive help', 'defective suppression', polyclonal activation, or idiotypic interactions. Induction of T cell-mediated autoimmunity by immunization with a protein of a different but related species, with haptenized, autologous proteins, or with autologous proteins associated with strong adjuvants have been interpreted to represent examples of the 'excessive help' concept. Defective 'suppression' that follows, *e.g.*, neonatal thymectomy, irradiation, cyclosporin treatment, or genetically determined defects of the apoptotic response favors autoimmunization of T cells. Polyclonal T cell activation may underly some autoimmune phenomena observed during chronic graft-*versus*-host reactions. T cell vaccination studies and some experimental arthritis models indicate the possibility of idiotype-specific regulation of T cell activation and suppression.

If it is accepted that self-reactive T cells arise and that they can potentially overcome or bypass the regulatory controls that usuallly prevent their activation, their clonal expansion and/or the expression of their effector functions, the questions of autoantigen processing and presentation gain central importance.

Processing pathways and presentation of autoantigens

Like all T cell-stimulating protein antigens, autoantigens have to be processed by specialized cells (APC, antigen-presenting cells), and epitopes derived from these antigens have to be presented on the cell surface in the context of major histocompatibility complex (MHC)-encoded glycoproteins. The term *'processing'* describes the partial proteolytic degradation of protein antigens in a relevant subcellular compartment to generate peptides that can bind in a haplotype-specific way to a particular class of integral membrane glycoproteins encoded within the MHC. Some factors are critical in processing. The selection of the type of APC that expresses or takes up the autoantigen, processes it, and presents its epitopes in the context of class I or class II MHC molecules determines if a specific, autoreactive T cell response is induced, and what kind of phenotype this response displays. The selection of alternative processing pathways available within a given type of APC, and the efficiency of processing within a given pathway, seem to be critical prerequisites for successful presentation. Furthermore, the 'costimulator profile' expressed by an APC type determines its ability to prime T cells. A complex sequence of events hence precedes immunogenic presentation of an autoantigen to self-reactive T cells. Within the last ten years we have learned to distinguish two alternative antigen processing pathways.

Processing of soluble protein antigens for MHC class II-restricted presentation to CD4$^+$ T cells

To enter the '*exogenous processing pathway*', soluble protein antigens in the extracellular fluid have to be endocytosed by the APC, or integral membrane proteins expressed by these cells on the surface have to be internalized.[2-4] Proteins taken up by the APC are denatured in an acid milieu and proteolytically degraded into peptidic fragments which bind to MHC class II molecules in a specialized endosomal compartment recently termed 'compartment for peptide loading' (CPL).[5-7] 'Loaded' class II molecules displayed on the cell surface are recognized by CD4$^+$ T cells. Hence, immunization with soluble protein antigens preferentially stimulates CD4$^+$ T cells. In most experimental models for T cell-mediated autoimmune diseases, CD4$^+$ T cells play a central role in the pathogenesis of the lesions. This has been convincingly demonstrated, *e.g.* for experimental allergic encephalitis[8-12] or adjuvant-induced arthritis.[13-20] In these systems, the relevant autoantigens (*i.e.* myelin basic protein, collagen or stress proteins) are processed in the exogenous pathway. Very few systems have been reported up to now in which class I-restricted CD8$^+$ T cells are the self-reactive effector cells of an experimentally induced autoimmune disease.

Processing of protein antigens for MHC class I-restricted presentation to CD8$^+$ T cells

Peptidic fragments generated from cytosolic, nuclear or mitochondrial protein antigens are degraded in the alternative '*endogenous processing pathway*' and stimulate class I-restricted CD8$^+$ T cells.[21-

[24] The large majority of peptides that stimulate CD8$^+$ T cells are produced from intracellular proteins by partial proteolytic degradation within the cytosol. The generated peptides are translocated across the membrane of the endoplasmic reticulum (ER) by a heterodimeric transporter complex formed by the two proteins TAP1 and TAP2. Haplotype-specific binding of peptides with a length of 8-10 residues to nascent MHC class I heavy chain/$\beta_2$microglobulin (β_2m) dimers takes place in the ER lumen and leads to transport-competent, trimeric complexes. This process seems to be rendered very efficient by the non-covalent association of the TAP dimer with the nascent MHC class I heavy chain/β_2m dimer in the ER membrane.[25] These stable trimers are rapidly conveyed by the default secretory route to the APC surface where they selectively stimulate CD8$^+$ T cells. This seems to be the major pathway for the generation and presentation of antigenic peptides by MHC class I molecules operating in many APC types under physiologic conditions.

Alternative processing pathways for MHC class I-restricted epitope presentation

Alternative pathways for loading antigenic peptides to MHC class I molecules have recently emerged. We can distinguish between *(i)* binding of nascent class I heavy chain/β_2m light chain to peptides transported into the ER lumen by alternative mechanisms, and *(ii)* 'loading' of antigenic peptides to class I molecules in an endosomal subcellular compartment. We will briefly review the available evidence for alternative mechanisms of the TAP-independent transport of antigenic peptides into the ER lumen, and will then focus the discussion on the binding of peptides processed from endocytosed, 'exogenous' protein antigens to MHC class I molecules in an endosomal compartment.

Integral membrane proteins with N-terminal leader sequences are cotranslationally translocated across the ER membrane. Signal sequences are cleaved off in the ER lumen by a signal peptidase. Peptides derived from signal sequences apparently bind to nascent MHC class I molecules because they have been eluted from class I molecules in normal and TAP-deficient cells.[26-30] Alternative carrier systems in the ER membrane may translocate proteins or peptides from the cytosol into the ER lumen. An attractive candidate would be stress proteins that bind proteins and translocate them through membranes,[31-33] but there is so far no experimental evidence to support a role of stress proteins in transporting peptides from the cytosol into the ER lumen for binding to MHC class I molecules. A rapid, retrograde transport of endocytosed shiga toxin into the ER lumen has been described.[34,35] This observation suggests that protein antigens or peptides endocytosed by the cell may be rapidly transported into the ER lumen where they may bind to MHC class I molecules. There is currently no experimental support for an 'exogenous' protein antigen to use this hypothetic pathway for class I restricted epitope presentation.

The data available up to now indicate that the ER lumen is the major subcellular compartment for the assembly of stable class I heavy chain/β_2m/peptide trimers under normal, physiologic conditions. Emerging evidence suggests that alternative processing pathways for class I-restricted epitope presentation operating in the endosomal compartment may be of equal importance, especially under pathologic conditions.

Peptide loading of MHC class I molecules in the endosomal compartment

The endosomal compartment is usually implicated in peptide loading of MHC class II molecules but not in peptide loading of MHC class I molecules. The first indication for an alternative pathway for peptide binding to MHC class I molecules was the observation that the injection of some soluble protein antigens efficiently primes class I-restricted cytotoxic T lymphocytes (CTL) *in vivo.* It is known that selected adjuvants, such as ISCOMs,[36-38] squalane/Tween 80,[39] saponins[40] or liposomes,[41-44] facilitate processing of 'exogenous' protein antigens for class I-restricted epitope presentation but the antigen is delivered into the cytosol of the presenting cell in at least some of these systems. Of interest are immunization experiments in which class I-restricted CTL were primed by 'exogenous' protein antigens injected without adjuvants. CTL priming by the injection of soluble protein antigen may be initiated by either phagocytosis or endocytosis of the antigen by the relevant APC.

Phagocytosis of bacteria results in presentation of peptides derived from bacterial proteins by class I MHC molecules.[45] This processing mechanism is resistant to cycloheximide and brefeldin A (*i.e.* does not involve nascent MHC class I molecules in the ER and their transport to the cell surface) but is sensitive to cytochalasin. After phagocytosis of the bacteria, antigenic peptides are 'regurgitated' by the processing cells that sensitize 'bystander' cells for class I-restricted lysis by CTL. This process hence resembles pulsing of cells for 10-20 minutes with synthetic, antigenic peptides of the appropriate length that sensitizes them for class I-restricted CTL lysis. This has been proposed to involve exchange of peptides of surface-associated MHC class I trimers.[22] The observation of successful class I-restricted epitope presentation after a brief *in vitro* pulse with synthetic peptides seen in fixed cells, in metabolically blocked cells or in cells kept at low temperatures is compatible with this notion. Many questions concerning the class I-restricted presentation of epitopes derived from phagocytosed protein antigens remain unanswered. The subcellular compartment (*e.g.* phagosomes or phagolysosomes) in which antigenic peptides are generated is not yet defined. It is uncertain if only regurgitated peptides bind to MHC class I molecules expressed on the surface of the processing cell or on the surface of a bystander in the immediate vicinity of the processing cell.

Endocytosis of protein antigen by APC leads to class I-restricted presentation of their epitopes. This was first suggested by data obtained in vaccination experiments. Injection of low doses of recombinant protein antigens without adjuvants by different routes efficiently primes class I-restricted CTL *in vivo*. This was demonstrated for influenza virus hemagglutinin,[46] influenza virus nucleoprotein,[47] the large T-antigen (T-Ag) of simian virus 40,[48-51] the surface protein (S-antigen) of hepatitis B virus (HBV),[52-56] denatured ovalbumin (OVA),[57] the denatured HIV-1 envelope protein gp160,[58] the HIV-1 nucleoprotein *gag*[59] and yeast-derived, hybrid Ty-virus-like particles (VLP).[60,61] The list of antigens tested in these systems includes native, particulate protein antigen constructs (HIV-1 *gag* particles, HBV surface particles, yeast Ty-VLP) as well as denatured, recombinant proteins (OVA, influenza virus nucleoprotein and hemagglutinin, denatured HIV-1 gp160 envelope protein). All these proteins are taken up by APC through endocytosis. In an *in vitro* system, we have investigated the class I-restricted presentation of an epitope processed from endocytosed HBV surface (S) antigen to murine, cloned CTL.[62] Efficient L^d-restricted presentation of the antigenic S_{28-39} epitope to CTL was evident in cells of different tissue origin that were pulsed *in vitro*, either with the antigenic S_{28-39}

S-peptide or with particulate S-antigen. Uptake of S-particles but not uptake of S-peptides by metabolically active APC was required for successful class I-restricted presentation of the antigenic epitope. Intracellular processing of S-particles was blocked by treating the APC with chloroquine, NH_4Cl or leupeptin. Processing and class I-restricted presentation were not blocked in cells treated with cytochalasin B or D, cycloheximide or brefeldin A. This processing pathway operated efficiently in peptide-transporter-deficient, L^d-transfected T2 cells. Sensitization of bystander cells by regurgitated peptides was not involved. These data revealed for the first time a novel endosomal/lysosomal processing pathway for class I-restricted presentation of peptides derived from endocytosed, 'exogenous' protein particles.

Stimulation of class I-restricted, autoreactive T cell responses by endocytosis of self-protein antigens

Injection (by various routes) of high doses of a 'foreign', immunogenic protein antigen in its native form into a mouse usually does not prime class I-restricted CTL; in contrast, injection of a 10^2- to 10^3-fold lower dose of the same antigen in its denatured form into the animal efficiently primes a class I-restricted CTL response. We have demonstrated this independently for two viral antigens, *i.e.* OVA[57] and HIV-1 gp160.[58] Although the details of the changes introduced by the denaturation process on the processing pattern of the antigen for class I-restricted epitope presentation are not yet available, we think this observation is relevant for the discussion of priming self-reactive, class I-restricted CTL. The TCR repertoire is selected during thymic education. Convincing evidence is available that many self-reactive class I- and class II-restricted T cell clones are deleted at this stage of T cell development to establish self tolerance. A prerequisite for this type of tolerance induction is the 'immunogenic' presentation (in the context of class I or class II MHC molecules) of peptides derived by processing of self proteins. Hence, only adequately processed self proteins can tolerize the T cell system. It is expected that during normal development it is only the native form of the self protein that will enter the processing machinery of thymic APC and hence contribute to the selection of the TCR repertoire. Hence, tolerance to peptides derived from alternative processing of denatured self proteins is not expected to be established in the thymus. It is therefore of interest to ask: is there experimental support for *the hypothesis that denatured self proteins can prime MHC-restricted T cells in vivo*?

We have obtained preliminary experimental evidence that a mechanism to bypass self-tolerance of class I-restricted T cells by allowing a self protein access to an alternative processing pathway through denaturation may in fact operate. These data were generated in our studies on self-reactive, class I-restricted CTL responses to determinants of the variable region of the β-chain of murine TCR. We studied the induction of self-reactive CTL responses to determinants of the variable Vß8.2 region of the TCR ß-chain in mice. Peptides were selected from the germline sequence of the Vß8.2 gene segments which contained putative K^b-, D^b- K^d- and L^d-restricted epitopes. Immunization of mice with some of these Vß8.2 peptides efficiently primed specific $CD8^+$ CTL responses *in vivo* against TCR-Vß8.2 epitopes generated during natural processing of this antigen.[63,64] This was demonstrated by testing *in vitro* the specific cytolytic reactivity of the *in vivo* peptide-primed, self-reactive CTL against syngeneic transfectants expressing a rearranged $Vß8.2^+$ TCR ß-chain. These transfectants were specifically lysed, demonstrating that TCR peptide vaccination induces self-reactive $CD8^+$ CTL that recognize naturally processed Vß8.2 epitopes. $Vß8.2^+$ T cells were not deleted in these TCR peptide-immunized mice, *i.e.* the fraction of $Vß8.2^+$ $CD4^+$ and $Vß8.2^+$ $CD8^+$ T cells in spleen and lymph nodes was not altered in vaccinated mice. The proliferative response of $Vß8.2^+$ T cells to stimulation by the mAb F23.2 was selectively suppressed in peptide-immunized mice, indicating partial anergy of this T subset. TCR peptide vaccination hence anergizes but does not delete most $Vß8.2^+$ T cells that present a naturally processed epitope of the TCR β-chain. The anti-TCR CTL may operate in Vβ-specific immunoregulation of the T cell system *in vivo*, suppressing their functional reactivity without deleting them. We obtained similar data when we immunized mice with a bacterially expressed, recombinant protein that represented the extracellular domains of a rearranged mouse $Vβ8.2^+$ TCR β-chain (A.K and J.R., unpublished). The self protein was in this case purified from its bacterial source using the detergent deoxycholate, and was therefore a denatured self protein. Hence, the non-physiological way chosen to introduce the autoantigen in a soluble, denatured form into the system (without adjuvants) may have introduced it into a non-physiologic processing pathway. This resulted in the immunogenic presentation of epitopes of a self protein against which the organism was not tolerant.

These data support the hypothesis that denatured self proteins gain access to alternative pathways for processing and MHC-restricted presentation of their epitopes. The cell biology and molecular biology underlying the observation that a particular type of denaturation of a protein antigen (*i.e.* detergent treatment) dramatically changes its immunogenicity for certain T cell subsets is not yet resolved. These findings have important implications for our concepts as to the induction of T cell-mediated autoaggressive reactions as well as for the design of vaccines that selectively stimulate certain T cell subsets.

Stress protein-mediated, endosomal processing of damaged cytosolic proteins for class I-restricted epitope presentation

It was pointed out above that denaturation of an 'exogenous' protein antigen dramatically enhances its potency to prime class I-restricted CTL by changing its processing pattern in the APC. In a different antigen system, we obtained evidence that damage of an intracellular (cytosolic or nuclear) protein antigen by, for example, mutation or truncation strikingly changes its processing pattern and hence the MHC class I-restricted presentation of immunogenic peptides derived from it. The system we developed is admittedly complex because it involves the expression of a panel of mutated viral genes in mutated cell lines that display defined defects of selected antigen processing pathways. The details of this system have been published.[65]

Murine RMA-S cells are deficient in translocating peptides from the cytosol into the lumen of the ER and therefore fail to present many intracellular protein antigens to CTL. Two class I (D^b)-restricted epitopes of simian virus 40 T-Ag were presented to CTL by wild-type wtT-Ag-transfected RBL5 or RMA cells but not by transfected RMA-S cells. Hence, 'endogenous' processing and presentation of the wtT-Ag depended on a functional peptide transporter heterodimer. In contrast, both T-Ag epitopes were efficiently presented to CTL by transfected RMA-S cells expressing a truncated, cytoplasmic T-Ag variant (cT-Ag) or a karyophilic, N-terminal T-

Ag fragment containing 272 amino acids. Transporter-independent 'endogenous' processing of mutant T-Ag molecules correlated with their association with the constitutively expressed heat shock protein 73 (hsp73). Hsp73 is known to bind misfolded proteins in the cytosol and to transport them into a lysosomal or endosomal compartment for degradation.[66-68] Efficient presentation of CTL-defined epitopes by RMA-S cells hence correlated with the association of these protein antigens to hsp73, indicating the existence of an alternative, peptide transporter-independent pathway for class I-restricted presentation of intracellular proteins. Class I-restricted presentation of both epitopes processed from these hsp73 associated protein antigens was sensitive to NH_4Cl and chloroquine. These data indicate that selected intracellular proteins access an alternative, hsp73-mediated pathway for class I-restricted presentation that operates independently of peptide transporters in an endosomal compartment.[65]

Concluding remarks

The T cell system is 'educated' to respond specifically to cell-associated epitopes of protein antigens presented in the context of different classes of MHC molecules. Processing, *i.e.* the partial proteolytic degradation of the antigen, precedes MHC-restricted presentation of epitopes to T cells. Alternative pathways of antigen processing operate in individual APC. The different types of antigen processing may yield different patterns of immunogenic peptides that are presented by MHC molecules to T cells. Under physiologic steady-state conditions, self proteins are introduced into the processing machinery in their native form. Hence, the T cell receptor repertoire in the healthy organism is expected to be tolerized during development to a spectrum of self peptides that result from processing of native autoantigens. Pathological stimuli can dramatically change the pattern of processing of a protein antigen; and a different spectrum of antigenic peptides can hence be generated from damaged (denatured) self proteins introduced into an alternative processing pathway. We have described experimental evidence that damaged 'exogenous' protein antigens (*i.e.* soluble, extracellular protein antigens denatured by detergents), as well as damaged 'endogenous' protein antigens (*i.e.* a truncated cytosolic protein) are processed in alternative pathways for class I-restricted presentation.

A stimulus that damages either a self protein or the biochemical integrity of an APC may thus prime self-reactive T cells. Further experimental support for this hypothesis is eagerly awaited.

Acknowledgement:

This work was supported by grants from the 'Deutsche Forschungsgemeinschaft' (Rei 549/4-1), the HERTIE Foundation and the FRITZ-THYSSEN-Foundation to J.R.

References

1 Schwartz RS. Autoimmunity and autoimmune disease. In: Paul WE. Fundamental Immunology. 3. New York: Raven Press, 1993:1033-98.
2 Brodsky FM, Guagliardi LE. The cell biology of antigen processing and presentation. Ann Rev Immunol 1991;9:707-44.
3 Germain RN, Hendrix LR. MHC class II structure, occupancy and surface expression determined by post-endoplasmic reticulum antigen binding. Nature 1991;353:134-9.
4 Neefjes JJ, Ploegh HL. Intracellular transport of MHC class II molecules. Immunol Today 1992;13:179-84.
5 Amigorena S, Drake JR, Webster P, Mellman I. Transient accumulation of new class II MHC molecules in a novel endocytic compartment in B lymphocytes. Nature 1994;369:113-20.
6 Tulp A, Verwoerd D, Dobberstein B, Ploegh HL, Pieters J. Isolation and characterization of the intracellular MHC class II compartment. Nature 1994;369:120-6.
7 West MA, Lucocq JM, Watts C. Antigen processing and class II MHC peptide-loading compartments in human B-lymphoblastoid cells. Nature 1994;369:147-51.
8 Sakai K, Sinha AA, Mitchell DJ, Zamvil SS, Rothbard JB, McDevitt HO, Steinman L. Involvement of distinct murine T-cell receptors in the autoimmune encephalitogenic response to nested epitopes of myelin basic protein. Proc Natl Acad Sci USA 1988;85:8608-12.
9 Sakai K, Zamvil SS, Mitchell DJ, Hodgkinson S, Rothbard JB, Steinman L. Prevention of experimental encephalomyelitis with peptides that block interaction of T cells with major histocompatibility complex proteins. Proc Natl Acad Sci USA 1989;86:9470-4.
10 Sun D, Klinkert WEF. Functional heterogeneity among $CD4^+$ encephalitogenic T cells in recruitment of $CD8^+$ T cells in experimental autoimmune encephalomyelitis. J Immunol 1989;143:2867-72.
11 Wraith DC, Smilek DE, Mitchell DJ, Steinman L, McDevitt HO. Antigen recognition in autoimmune encephalomyelitis and the potential for peptide-mediated immunotherapie. Cell 1989;59:247-55.
12 Gaur A, Wiers B, Liu AY, Rothbard JA, Fathman CG. Amelioration of au-

toimmune encephalomyelitis by myelin basic protein synthetic peptide-induced anergy. Science 1992;258:1491-4.

13 Van Eden W, Holoshitz J, Nevo Z, Frenkel A, Klajman A, Cohen IR. Arthritis induced by a T-lymphocyte clone that responds to Mycobacterium tuberculosis and to cartilage proteoglycans. Proc Natl Acad Sci USA 1985;85:5117-20.

14 Hom JT, Stuart JM, Chiller JM. Murine T cells reactive to type II collagen. I. Isolation of lines and clones and characterization of their antigen-induced proliferative response. J Immunol 1986;136:769-75.

15 Cooper SM, Sriram S, Ranges GE. Suppression of murine collagen-induced arthritis with monoclonal anti-Ia antibodies and augmentation with IFN-γ. J Immunol 1988;141:1958-62.

16 Seki N, Sudo Y, Yoshioka T, Sugihara S, Fujitsu T, Sakuma S, Ogawa T, Hamaoka S, Senoh H, Fujiwara H. Type II collagen-induced murine arthritis. I. Induction and perpetuation of arthritis require synergy between humoral and cell-mediated immunity. J Immunol 1988;140:1477-84.

17 Van Eden W, Thole JER, Van der Zee R, Noordzij A, Van Embden JDA, Hensen EJ, Cohen IR. Cloning of the mycobacterial epitope recognized by T lymphocytes in adjuvant arthritis. Nature 1988;331:171-3.

18 DeJoy SQ, Ferguson KM, Sapp TM, Zabriskie JB, Oronsky AL, Kerwar SS. Streptococcal cell wall arthritis. Passive transfer of disease with a T cell line and crossreactivity of streptococcal cell wall antigens with Mycobacterium tuberculosis. J Exp Med 1989;170:369-82.

19 Murray JS, Madri J, Tite J, Carding SR, Bottomly K. MHC control of $CD4^+$ T cell subset activation. J Exp Med 1989;170:2135-40.

20 Hermann E, Lohse AW, Van der Zee R, Van Eden W, Mayet WJ, Probst P, Poralla T, Meyer zum Büschenfelde K-M, Fleischer B. Synovial fluid-derived *Yersinia*-reactive T cells responding to human 65-kDa heat-shock protein and heat-stressed antigen-presenting cells. Eur J Immunol 1991;21:2139-43.

21 Townsend ARM, Bodmer HC. Antigen recognition by class I-restricted T lymphocytes. Ann Rev Immunol 1989;7:601-24.

22 Yewdell JW, Bennink JR. Cell biology of antigen processing and presentation to major histocompatibility complex class I molecule-restricted T lymphocytes. Adv Immunol 1992;52:1-123.

23 Germain RN, Margulies DH. The biochemistry and cell biology of antigen processing and presentation. Ann Rev Immunol 1993;11:403-50.

24 Rammensee H-G, Falk K, Rötzschke O. Peptides naturally presented by MHC class I molecules. Ann Rev Immunol 1993;11:213-44.

25 Ortmann B, Androlewicz MJ, Cresswell P. MHC class I/β_2-microglobulin complexes associate with TAP transporters before peptide binding. Nature 1994;368:864-7.

26 Henderson RA, Michel H, Sakaguchi K, Shabanowitz J, Appella E, Hunt DF, Engelhard VH. HLA-A2.1-associated peptides from a mutant cell line: A second pathway of antigen presentation. Science 1992;255:1264-8.

27 Hunt DF, Henderson RA, Shabanowitz J, Sakaguchi K, Michel H, Sevilir N, Cox AL, Appella E, Engelhard VH. Characterization of peptides bound to the class I MHC molecule HLA-A2.1 by mass spectrometry. Science 1992;255:1261-3.

28 Wei ML, Cresswell P. HLA-A2 molecules in an antigen-processing mutant cell contain signal sequence-derived peptides. Nature 1992;356:443-6.
29 Huczko EL, Bodnar WM, Benjamin D, Sakaguchi K, Zhu NZ, Shabanowitz J, Henderson RA, Appella E, Hunt DF, Engelhard VH Characteristics of endogenous peptides eluted from the class I MHC molecule HLA-B7 determined by mass spectrometry and computer modeling. J Immunol 1993;151:2572-88.
30 Engelhard VH. Structure of peptides associated with class I and class II MHC molecules. Ann Rev Immunol 1994;12:181-207.
31 Craig EA, Gambill BD, Nelson RJ. Heat shock proteins: Molecular chaperones of protein biogenesis. Microbiol Rev 1993;57:402-14.
32 Craig EA. Chaperones: Helpers along the pathways to protein folding. Science 1993;260:1902-3.
33 Welch WJ, Suhan JP. Cellular and biochemical events in mammalian cells during and after recovery from physiological stress. J Cell Biol 1993;103:2035-52.
34 Sandvig K, Garred O, Prydz K, Kozlov JV, Hansen SH, Van Deurs B. Retrograde transport of endocytosed Shiga toxin to the endoplasmic reticulum. Nature 1992;358:510-2.
35 O'Brien AD, Tesh VL, Donohue-Rolfe A, Jackson MP, Olsnes S, Sandvig K, Lindberg AA, Keusch GT. Shiga toxin: Biochemistry, genetics, mode of action, and role in pathogenesis. Curr Top Microbiol Immunol 1992; 180:65-94.
36 Takahashi H, Takeshita T, Morein B, Putney S, Germain RN, Berzofsky JA. Induction of CD8$^+$ cytotoxic T cells by immunization with purified HIV-1 envelope protein in ISCOMs. Nature 1990;344:873-5.
37 Heeg K, Kuon W, Wagner H. Vaccination of class I major histocompatibility complex (MHC)-restricted murine CD8$^+$ cytotoxic T lymphocytes towards soluble antigens: immunostimulating-ovalbumin complexes enter the class I MHC-restricted antigen pathway and allow sensitization against the immunodominant peptide. Eur J Immunol 1991;21:1523-7.
38 Van Binnendijk RS, Van Baalen CA, Poelen MCM, De Vries RRP, Boes J, Cerundolo V, Osterhaus ADME, Uytdehaag FGCM. Measles virus transmembrane fusion protein synthesized *de novo* or presented in immunostimulating complexes is endogenously processed for HLA class I- and class II-restricted cytotoxic T cell recognition. J Exp Med 1992;176:119-28.
39 Raychaudhuri S, Tonks M, Carbone F, Ryskamp T, Morrow WJ, Hanna N. Induction of antigen-specific class I-restricted cytotoxic T cells by soluble proteins *in vivo.* Proc Natl Acad Sci USA 1992;89:8308-12.
40 Newman MJ, Wu JY, Gardner BH, Munroe KJ, Leombruno D, Recchia J, Kensil CR, Coughlin RT. Saponin adjuvant induction of ovalbumin-specific CD8$^+$ cytotoxic T lymphocyte responses. J Immunol 1992;148:2357-62.
41 Huang L, Reddy R, Nair SK, Zhou F, Rouse BT. Liposomal delivery of soluble protein antigens for class I MHC-mediated antigen presentation. Res Immunol 1992;143:192-6.
42 Nair S, Zhou X, Huang L, Rouse BT. Class I restricted CTL recognition of a soluble protein delivered by liposomes containing lipophilic polylysines. J Immunol Methods 1992;152:237-43.

43 Reddy R, Zhou F, Nair S, Huang L, Rouse BT. *In vivo* cytotoxic T lymphocyte induction with soluble proteins administered in liposomes. J Immunol 1992;148:1585-9.
44 Zhou F, Rouse BT, Huang L. Induction of cytotoxic T lymphocytes *in vivo* with protein antigen entrapped in membranous vehicles. J Immunol 1992;149:1599-1604.
45 Pfeifer JD, Wick MJ, Roberts RL, Findlay K, Normark SJ, Harding CV. Phagocytic processing of bacterial antigens for class I MHC presentation to T cells. Nature 1993;361:359-62.
46 Yamada A, Ziese MR, Young JF, Yamada Y, Ennis FA. Influenza virus hemagglutinin-specific cytotoxic T cell response induced by polypeptide produced in Escherichia coli. J Exp Med 1985;162:663-74.
47 Wraith DC, Vessey AE, Askonas BA. Purified influenza virus nucleoprotein protects mice from lethal infection. J Gen Virol 1987;68:433-40.
48 Tevethia SS, Flyer DC, Tjian R. Biology of simian virus 40 (SV40) transplantation antigen(TrAg). VI. Mechanism of induction of SV40 transplantation immunity in mice by purified SV40 T antigen(D2 protein). Virology 1980;107:13-23.
49 Schirmbeck R, Zerrahn J, Kuhröber A, Kury E, Deppert W, Reimann J. Immunization with soluble simian virus 40 large T antigen induces a specific response of $CD3^+$ $CD4^-CD8^+$ cytotoxic T lymphocytes in mice. Eur J Immunol 1992;22:759-66.
50 Schirmbeck R, Deppert W, Kury E, Reimann J. *In vitro* expansion distorts the detectable pattern of specific recognition of *in vivo* primed cytotoxic T lymphocyte populations. Cell Immunol 1993;149:444-9.
51 Schirmbeck R, Zerrahn J, Kuhröber A, Deppert W, Reimann J. Immunization of mice with the N-terminal (1-272) fragment of simian virus 40 large T-antigen (without adjuvants) specifically primes cytotoxic T lymphocytes. Eur J Immunol 1993;23:1528-34.
52 Barnaba V, Franco A, Alberti A, Balsano C, Benvenuto R, Balsano F. Recognition of hepatitis B virus envelope proteins by liver infiltrating T lymphocytes in chronic HBV infection. J Immunol 1989;143:2650-5.
53 Barnaba V, Franco A, Alberti A, Benvenuto R, Balsano F. Selective killing of hepatitis B envelope antigen-specific B cells by class I-restricted, exogenous antigen-specific T lymphocytes. Nature 1990;345:258-60.
54 Schirmbeck R, Melber K, Mertens T, Reimann J. Selective stimulation of murine cytotoxic T cell and antibody responses by particulate or monomeric hepatitis B virus surface (S) antigen. Eur J Immunol 1994;24:1088-96.
55 Schirmbeck R, Melber K, Mertens T, Reimann J. Antibody and cytotoxic T-cell responses to soluble hepatitis B virus (HBV) S antigen in mice: implication for the pathogenesis of HBV-induced hepatitis. J Virol 1994;68:1418-25.
56 Schirmbeck R, Melber K, Kuhröber A, Janowicz ZA, Reimann J. Immunization with soluble hepatitis B virus surface (S) protein particles elicits murine H-2 class I-restricted $CD8^+$ cytotoxic T lymphocyte responses *in vivo.* J Immunol 1994;152:1110-9.
57 Schirmbeck R, Böhm W, Reimann J. Injection of detergent-denatured ovalbumin primes murine class I-restricted cytotoxic T cells *in vivo.* Eur J Immunol

1994;24:2068-72.
58 Schirmbeck R, Deml L, Melber K, Wolf H, Wagner R, Reimann J. Priming of class I-restricted cytotoxic T lymphocytes by vaccination with recombinant protein antigens. Vaccine 1994;13:857-65.
59 Griffiths JC, Harris SJ, Layton GT, Berrie EL, French TJ, Burns NR, Adams SE, Kingsman AJ. Hybrid human immunodeficiency virus Gag particles as an antigen carrier system: induction of cytotoxic T-cell and humoral responses by a Gag:V3 fusion. J Virol 1993;67:3191-8.
60 Harris SJ, Gearing AJ, Layton GT, Adams SE, Kingsman AJ. Enhanced proliferative cellular responses to HIV-1 V3 peptide and gp120 following immunization with V3:Ty virus-like particles. Immunology 1992;77:315-21.
61 Layton GT, Harris SJ, Gearing AJ, Hill Perkins M, Cole JS, Griffiths JC, Burns NR, Kingsman AJ, Adams SE. Induction of HIV-specific cytotoxic T lymphocytes *in vivo* with hybrid HIV-1 V3:Ty-virus-like particles. J Immunol 1993;151:1097-1107.
62 Schirmbeck R, Melber K, Reimann J. Hepatitis B virus small surface particles are processed in a novel endosomal pathway for MHC class I-restricted epitope presentation. Eur J Immunol 1995;25:1063-70.
63 Kuhröber A, Schirmbeck R, Reimann J. Vaccination with T cell receptor peptides primes anti-receptor cytotoxic T lymphocytes (CTL) and anergizes T cells specifically recognized by these CTL. Eur J Immunol 1994;24:1172-80.
64 Kuhröber A, Schirmbeck R, Reimann J. A selfreactive class I-restricted T cell response of H-2^b mice to determinants of the Vβ8.2 domain of the T cell receptor for antigen. Immunology 1994;83:532-9.
65 Schirmbeck R, Reimann J. Peptide transporter-independent, stress protein-mediated endosomal processing of endogenous protein antigens for major histocompatibility complex class I presentation. Eur J Immunol 1994;24:1478-86.
66 Chiang H-L, Terlecky SR, Plant CP, Dice JF. A role for a 70-kilodalton heat shock protein in lysosomal degradation of intracellular proteins. Science 1989;246:382-5.
67 Dice JF. Peptide sequences that target cytosolic proteins for lysosomal degradation. Trends Biochem Sci 1990;15:305-9.
68 Terlecky SR, Chiang H-L, Olson TS, Dice JF. Protein and peptide binding and stimulation of in vitro lysosomal proteolysis by the 73-kDa heat shock cognate protein. J Biol Chem 1992;267:9202.

INGA MELCHERS

B cell response in autoimmunity

Autoimmunity may be pathogenic, associated with disease, or normal

The textbook definition is simple: "Diseases in which the pathology is caused by immune responses to self antigens are called autoimmune diseases."[1] It is more difficult to prove that a given disease is caused by an autoimmune response. Today we know that lymphocytes with receptors recognizing self antigens are common, and usually not pathogenic.[2–5] In order to understand the mechanisms leading to disease and the role of the immune system in this process, we have to proceed in two steps. First, autoimmune responses at the T or B cell level which are either normal or typical for a disease must be described. Then, the pathogenicity of the disease-associated responses must be proven.

A large number of diseases is known in which typical autoantibodies are detected in the sera of patients.[5,6] Frequently, these antibodies themselves are not pathogenic. Nevertheless it is worthwhile to study how these characteristic autoantibodies originate. Obviously, the answer might contribute to our understanding of the onset of the disease itself. Antibody secretion depends on the previous antigen-specific activation of a B cell, in the presence of T cell help.[7] Help is normally provided by a $CD4^+$ helper T cell, which in turn was previously specifically activated. Most helper T cells recognize peptide antigens bound to a MHC class II molecule on the surface of an antigen-presenting cell.[7] B cells not only bind antigens via their immunoglobulin receptors; in addition, they internalize, process and present antigens to T cells, which leads to the cooperation of both cell types in antibody production.[7] As human helper T cells are usually restricted to HLA class II antigens (DR, DP, DQ), it is not too surprising to find many hypothesized and proven autoimmune diseases associated with certain HLA haplotypes.[5] One might place B cells at the center of an autoimmune response, because they produce the autoantibodies and express the antigen-presenting HLA molecules – without implying that B cells themselves induce such responses (Fig. 1). Frequently, the peptide bound to the class II antigen is derived from the antigen that bound to the B cell receptor (BcR). However, this is not a necessary condition; B cells receiving signals via their BcR may be activated by T cells recognizing a different peptide bound to the B cell's class II molecule.

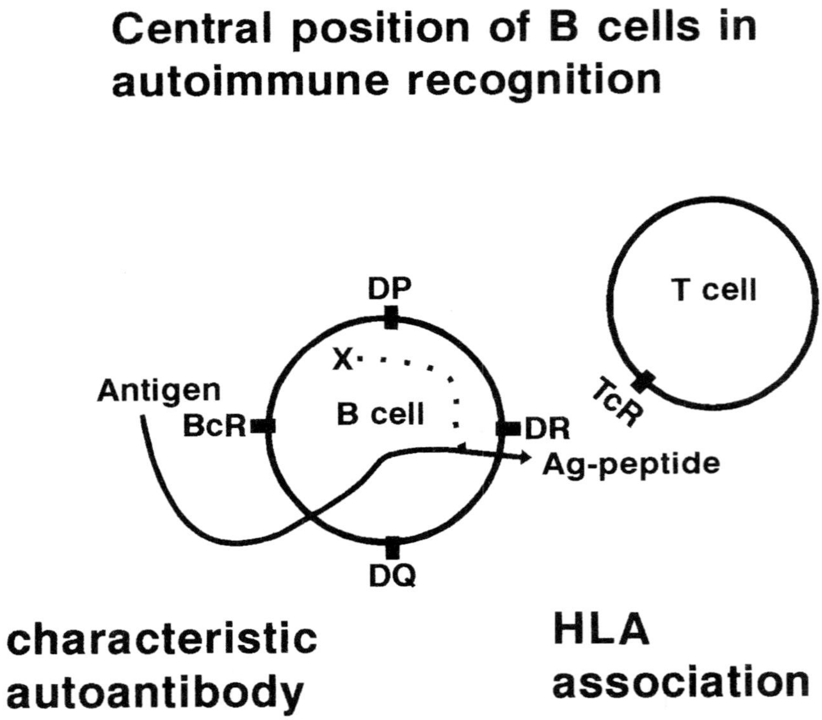

FIGURE 1 A simple model of the B cell interaction with antigen and an antigen-specific helper T cell. B cells express an antigen-specific B cell receptor (BcR) and HLA class II molecules (DP, DR, DQ) on their surface. The antigen binds to the BcR, is internalized, processed and eventually displayed as a peptide on a class II molecule, where it may be recognized in a restricted way by a T cell receptor (TcR). A TcR may also recognize a peptide derived from another protein X. The B cell receptors may represent characteristic autoantibodies; the class II genes may be the ones which are associated with the autoimmune disease.

The dynamic nature of the B cell population

B cells are generated from bone marrow stem cells throughout the life of an individual – probably 10^{10} per day.[8] B cell differentiation comprises many events, of which only some are mentioned in Fig. 2. The individual cell will rearrange its Ig-genes (V_H, D_H, J_H, C_H for the heavy chain and V_L, J_L, C_L for the light chain) to express only one Ig-receptor type on its surface. In man, the population of mature B cells consists of ca. 10^{12} cells with a large number of different antigen-specific receptor types. The expressed repertoire is a reflection of the germ line repertoire, but not identical with it. Both negative and positive selection processes shape the expressed B cell repertoire according to the environment of the individual immune system. These dynamic changes concern all mature B cells, and result in the extinction or expansion of individual clones (clonal selection; selected repertoire).[9] Every day, approximately 1% of the total B cell population dies and is replaced by a new generation. B cell activation mainly occurs in the lymphoid follicles and germinal centers of peripheral lymph nodes, spleen and gut-associated tissues. Some B cells pass through all stages of B cell differentiation and become plasma cells or memory cells. Some have only a short life, others return to the bone marrow and persist there for years. Following B-cell activation, mutations within the variable parts of the receptors

The individual B cell	
does	**may be**
differentiate	a naive precursor B cell
divide	
die	
rearrange Immunoglobulin genes	
express only one Ig receptor type	a mature B cell
secrete antibodies	an antibody-secreting cell (ASC), a plasma cell
migrate	
present antigen to T cells on MHC molecules class I and class II	an antigen-presenting cell
interact with T cells and others	

FIGURE 2 Some events in the life of a B cell.

may occur and the constant regions may switch to other isotypes. B cells with these newly acquired receptors also underlie selection processes and participate in creating the actual repertoire.

The large number of possibilities to form and change a B cell receptor, combined with clonal selection and the dynamics of cell growth and death, allows the immune system to be adaptable to changes in its environment. This is advantageous for the individual, as it might be the only way to fight microorganisms, which grow faster – and thus change faster – than a mammal does, effectively. However, it is also dangerous, as among the changing receptors some may appear which are autoreactive in addition to being antigen-reactive. Supplementary to selection processes, regulation is required to balance necessary antigen-reactivity and unsupportable autoreactivity.[10-12]

These considerations lead to many questions, such as the following: How does the immune system of an individual healthy organism distinguish self from non-self? How much and what type of autoimmunity is allowed or necessary? Where is the difference between normal autoimmunity and autoimmune disease?

How to study the B cell repertoire.

Many human autoantibodies have been known for a long time.[5] During the last 10 years, a still increasing number of autoantigens has been defined molecularly using the methodology of modern molecular biology.[5] Similarly, refined techniques are now available to study the cellular and molecular biology of B lymphocytes, both as individual cells and at the level of the population as a whole. It has become possible to study the rearranged Ig-genes of single cells in situ, at sites where autoantibodies are produced.[13,14] We and others immortalize B cells by virus transformation or cell fusion, and then study the features of monoclonal antibodies at the protein level.[15-20] The combination of an efficient polyclonal activation with limiting dilution analysis allows one to measure frequencies and specificities of antibody-secreting cells (ASC) and their precursors (p-ASC) at the population level.[21]

Fig. 3 depicts the experimental design originally published by Wen et al.,[22] which we used to gather the data described below.[23-26] In principle, we use a mutant murine lymphoma line to activate up to 50-100% of human B cells via CD40/CD40 ligand.[27] Fig. 4 de-

B cell activation in limiting dilution culture

1. **Enrich $CD19^+$ B cells from peripheral blood by depletion of T cells.**

2. **Place B cells into cultures in concentrations ranging from**
   A. 32 - 0,5 B cells/culture
   B. 1000 - 8 B cells/culture.
   Use ≥ 24 replicates per concentration.

3. **Add 5 x 10^4 irridiated EL-4 B5 cells/culture (mutant mouse T cell lymphoma), 10 % conditioned medium (containing cytokines), 3 ng/ml PMA.**

4. **Take supernatants 10 days later, and test them for the presence of antibodies, as total Ig (A) or specific Ig (B). Several assays may be performed with single supernatants, and the data correlated (e. g. for segregation analyses).**

5. **Calculate the fractions of negative cultures per concentration (F0), and perform Poisson analyses to estimate the frequencies of Ig-secreting cells or p-ASC for specific antibodies.**

6. **At low concentrations of B cells/culture segregation analyses can be performed which allow to analyse the specificity of individual clones.**

FIGURE 3 The experimental design to study the frequency and specificity of p-ASC at the population level. Further details are published in Rudolphi et al.[24]

picts an example of the Poisson analysis used to estimate frequencies, and the percentages of peripheral B cells, which we are able to activate from healthy donors or patients with rheumatoid arthritis (RA) or systemic lupus erythematosus (SLE).[23,24]

Differences between the normal and the autoimmune repertoire

SYSTEMIC LUPUS ERYTHEMATOSUS The clinical manifestations of SLE are arthritis, rash and glomerulonephritis; autoimmune thrombocytopenia, hemolytic anemia and involvement of the central nervous system are common complications. Many autoantibodies occur in SLE, some of which certainly possess pathogenic features (e.g., antibodies to thrombocytes). The most prominent autoantibodies are the IgGs binding to double-stranded DNA (dsDNA) with rela-

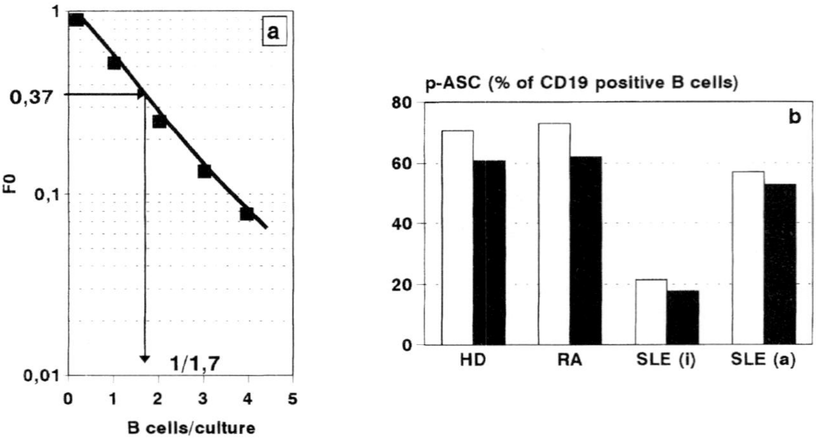

FIGURE 4 Efficiency of B cell activation. Peripheral CD19^{+} B cells were activated as described in Fig. 3. Total Ig (white bars) and IgM (black symbols) were measured in the supernatants of individual cultures, and the fractions of negative cultures calculated (Fo). The frequency of activated cells was estimated by Poisson analysis as demonstrated for one example (A). Activation is summarized in (B) as mean values calculated from a total of 39 individuals tested (11, 10, 7 and 11 donors per group, respectively). (I) = inactive, (a) = active.

tive high affinity. Since the serum levels of these autoantibodies correlate with disease activity, and depositions of DNA/anti-DNA immune complexes in the glomeruli of the kidney are associated with nephritis, these autoantibodies play a direct pathogenic role.[28,29] Analysis of hybridomas has shown that they use V-genes, revealing the characteristics of an antigen-driven selection and clonal expansion of autoreactive B cells.[30-32] Since, on the other hand, the B cells of SLE patients show features of polyclonally activated cells, the driving forces for this B cell expansion at the population level are still under discussion.

We estimated the frequencies of precursor cells (p-ASC) for Ig-anti-DNA in six healthy controls and 18 SLE patients with active and inactive disease (Fig. 5).[23] P-ASC for IgG-anti-dsDNA were exclusively detected in SLE patients (in 73% of those with active disease and in one with inactive disease). Their frequencies (0.01-0.99% of p-ASC for total IgG) were in the same order of magnitude as the frequencies of p-ASC for IgG-anti-tetanus-toxoid (TT), which were detectable in three healthy volunteers after booster vaccination (0.07-0.8% of p-ASC for total IgG), but not before ($\leq$

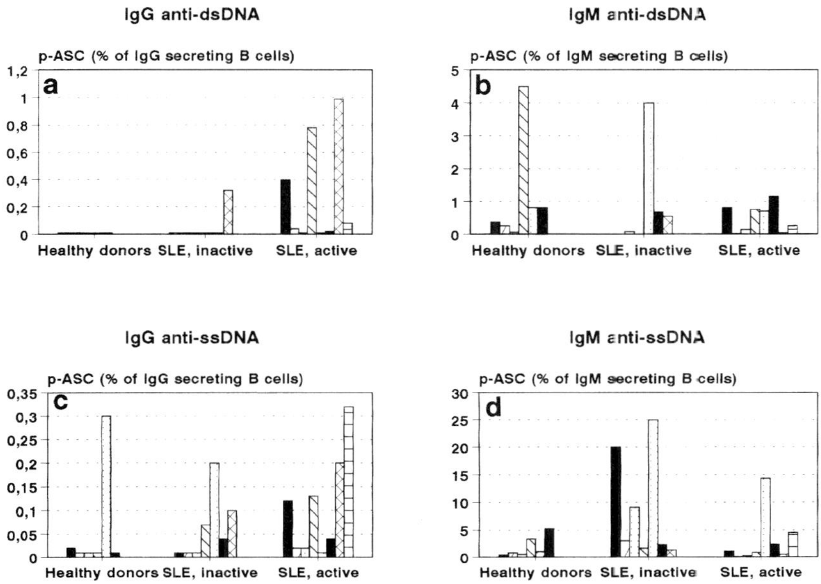

FIGURE 5 p-ASC specific for DNA in SLE patients and healthy blood donors. The frequencies obtained in individuals are expressed as percentages of p-ASC for IgG or IgM, respectively, which could be activated into differentiation. Experimental details are described in Schwab et al.[23]

0.01%). P-ASC for IgG-anti-ssDNA were observed in 33% of the healthy donors and in 78% of SLE patients in frequencies of 0.01–0.32% of p-ASC for total IgG. Only patient-derived IgG-anti-DNA clones cross-reacted with (33%) or were monoreactive to dsDNA (12%). P-ASC for IgM-anti-DNA were seen in healthy donors and SLE patients in comparable frequencies and with similar reactivities with ssDNA and dsDNA. These results indicate that precursor B cells for IgG-anti-dsDNA in SLE patients are similarly selected and expanded as are precursor B cells specific for foreign antigens such as TT.

RHEUMATOID ARTHRITIS Rheumatoid arthritis (RA) is a prevalent, chronic disease with inflammatory arthritis, usually symmetric affecting several to many joints, and additional complications like vasculitis. Rheumatoid factors (RF) are autoantibodies against the F_C-region of IgG molecules, present in the sera of most adult patients with RA (seropositive RA).[33] RF may be IgM, IgG or IgA. They are used as a criterium in the diagnosis of the disease.[34] However, RF also occur regularly in healthy individuals. They increase in number during infections and immune responses in general, and are thought to be important as cross-linking agents for the elimination of antigen-antibody complexes. B cells with RF as receptors take up antigen-antibody complexes, and can thus present antigens to T cells for which they are not specific themselves.[35] Most of these "naturally" occurring RF are thought to be multireactive and of low affinity, which is suitable for a first-line defense mechanism.

During the last few years, evidence has been collected which indicates that the RF of RA patients is not the same as the usual RF, but of a different specificity, coded for by different V-genes and equipped with more mutations.[36,37] Our own studies support the hypothesis that in RA patients, a specific B cell population has expanded which carries IgM receptors with a higher specificity for IgG F_C-fragments. These data are summarized in Fig. 6 and described in detail in Rudolphi et al.[24] The central observation is that in most seropositive RA patients we detect more p-ASC for monospecific RF than for multireactive RF.

All blood samples tested contained p-ASC for IgM RF in relative high frequencies. In 14 patients with seroposive RA and 8 healthy donors the mean frequencies of $CD19^+$ B cells were 1.48% and 0.91%. Although patients tend to have more of these p-ASC, this difference was not significant. However, when we looked at the specificities of these precursors, a significant difference was detected in the ratios of the frequencies for monospecific (S) and multireactive (M) p-ASC for RF (Fig. 6). In most seropositive RA patients the frequencies of p-ASC for monospecific IgM RF (B cell population S) were higher than the frequencies of p-ASC for multireactive IgM RF (B cell population M), whereas in healthy donors we always found the opposite – i.e., the frequencies of p-ASC for multireactive IgM RF were higher than the frequencies of p-ASC for monospecific IgM RF. This expansion of population S in seropositive RA patients seems to be disease specific. So far, we have found it in most – although not all – patients in this category, but have failed to see it in patients with seronegative RA, juvenile chronic arthritis, or reactive arthritis, nor in non-affected family members of RA patients.[25] We have seen it not only in the blood but also in synovial B cells and splenic B cells of a patient with Felty's Syndrome.[25]

What is the reason for the disease-specific expansion of the MS

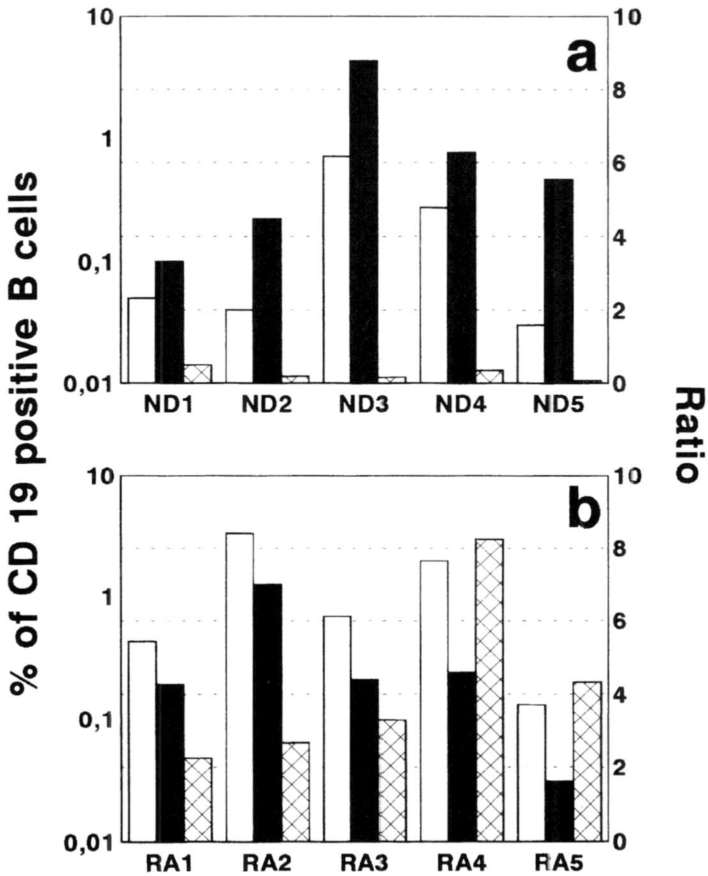

FIGURE 6 p-ASC for monospecific or multireactive rheumatoid factors in patients with rheumatoid arthritis and healthy blood donors. The frequencies obtained in individuals are expressed as percentages of $CD19^+$ B cells. In brief, culture supernatants were tested for IgM binding in ELISAs to human IgG F_C-fragments (= rheumatoid factors), human collagen types I, II, IV and V, and the mycobacterial stress protein hsp60. Only cultures from groups with Fo > 0.37 were used for this analysis. Rheumatoid factors were classified as monospecific (white bars) if they only reacted with human IgG F_C-fragments, and as cross-reactive or multireactive (black bars) if they reacted with at least one of the other antigens tested. In addition to frequencies, the ratio of the two frequencies is given for each individual (cross-hatched bars). In these experiments, the activation efficiency was 100%. Experimental details are described in Rudolphi et al.[24]

population in seropositive RA patients? It may be an antigen-driven process, where the antigen is unknown. There may also be other genes involved which are important for the formation of the B cell repertoire. In our earlier work with the immune response of mice to sheep erythrocytes, we observed that in addition to the MHC complex, genes within the V-gene region of the IgH-locus also determined the frequencies of p-ASC for anti-SRBC antibodies.[38] We therefore analyzed the distribution of IgV_H-haplotypes among seropositive RA patients and healthy controls in Europe in comparison to the same two groups in Japan.[39] As shown in Fig. 7, there is an interesting difference between the European and the Japanese population in general, but on both continents the distribution of IgV_H-haplotypes was the same in patients and in healthy donors.

Differences between the peripheral and the local repertoire

Blood contains B lymphocytes migrating from bone marrow into the peripheral lymphoid organs, or from there via the thoracic duct back into the recirculation and the bone marrow. B cells also reside in

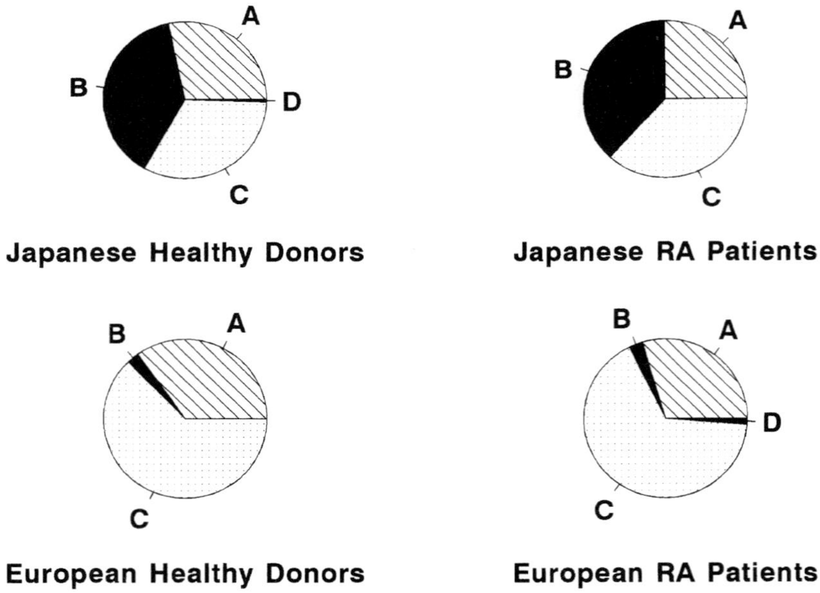

FIGURE 7 Distribution of 4 IgV_H–haplotypes (A,B,C,D) within different populations. The IgV_H- haplotypes and details of the analysis are described in Shin et al.[39] In brief, genomic DNAs were prepared from peripheral blood leukocytes of 181 healthy individuals randomly selected and from 134 patients with clinically defined RA and digested with restriction endonucleases. After electrophoresis and transfer of DNA to nitrocellulose filters, blots were hybridized with P79 and P4.1 – two probes isolated from the human V_H locus.

other local environments, e. g. synovial tissue and synovial fluid. Whereas many naive B cells are encountered in the blood stream, B cells in other compartments, especially in the affected organs of patients with autoimmune diseases, might be expected to represent a population enriched for cells that recognize relevant – local – autoantigens. Indeed, most peripheral B cells carry IgM receptors; in contrast, most synovial B cells have switched to IgG or IgA receptors.[25,26,40-42]

We were interested to find out whether we could detect higher frequencies of p-ASC with specificities for antigens that induce arthritis in animal models, like collagen type II (CII) or the stress protein hsp65 of *M. tub.*, in the synovial population.[43,44] Since the activation efficiency of synovial B cells differs from that of peripheral B cells, we calculated all frequencies on the basis of $CD19^+$ B cells which could be activated to secrete IgM, IgG or IgA.[42] We chose TT as a control antigen and found p-ASC for IgG- or IgA-anti-TT to be less frequent in the joint than in the blood ($[FS/FB] - 1 < 0$, where FS = frequency in the synovial population and FB = frequency in the blood; see Fig. 8A). We then estimated the frequencies of p-ASC for IgG- and IgA-anti-mycobacterial-hsp65, -human-CII, or -human-F_C-fragments. P-ASC for anti-hsp65 were enhanced more than 20-fold in 4 samples out of 9 tested (Fig. 8B), and p-ASC for anti-CII and p-ASC for IgA rheumatoid factors were enhanced in 2 (or 3) samples each (Figs. 8C and 8D). In the 9 patients tested with these three antigens, more than 90% of the repertoire of all p-ASC for IgG or IgA was missed, i.e. most B cells in these patients recognized other antigens.

Conclusions

All studies of the peripheral repertoire performed up to now have shown that a large percentage of B cells has the capacity to recognize autoantigens, both in patients with autoimmune disease and in healthy donors. However, there are also important differences between the two groups. In autoimmune patients, potentially autoreactive B cells are easily observed which have switched from IgM to other Ig-classes, and thus have been activated before. In SLE, the appearance of this population in the blood accompanies an active phase of the disease. The fine-specificity of the B cell receptors also differs between patients and healthy donors. In RA patients, this is

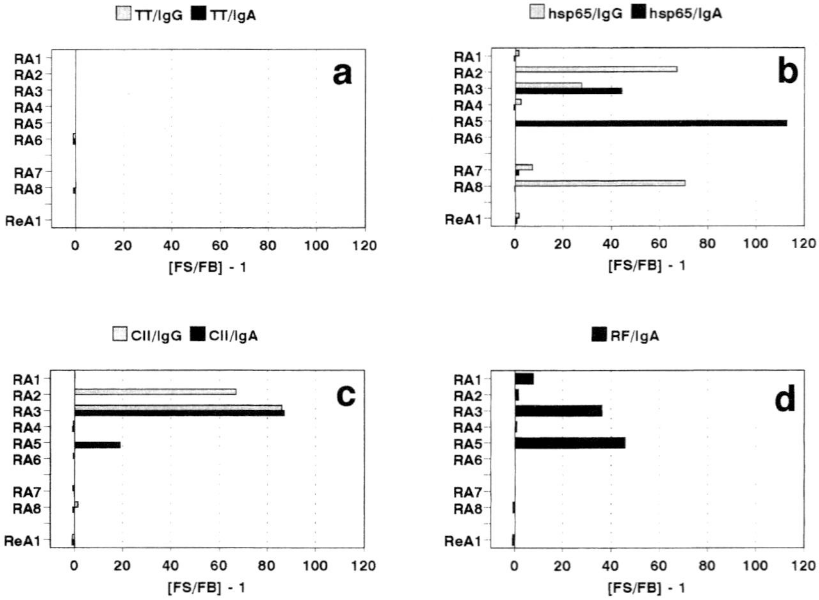

FIGURE 8 Comparison of the peripheral and the synovial B cell repertoire in patients with rheumatoid arthritis. B cells were prepared and analysed as described before.[24] Frequencies of p-ASC for IgG or IgA anti TT (A), hsp60 of *M. tub.* (B), human collagen type II (C) or human F_C-fragments (D) were estimated in both compartments. Frequencies of p-ASC within the synovial population (FS) and the peripheral blood population (FB) were then used to calculate the ratio [FS/FB]-1, which best shows an existing relative increase of specific p-ASC in the synovial population.

already visible at the level of IgM-bearing B cells, which are more frequently monospecific for IgG-F_C-fragments. In both cases, it is still unclear what induces or allows the expansion of these characteristic B cell populations. Much more needs to be learned about the human autoimmune and antigen-reactive repertoire and the movements within the immune system, which normally recognizes its own body without destroying it.

Summary

The B cell repertoire of healthy donors and of patients with autoimmune diseases has been analyzed at the level of the lymphocyte population. In both groups, a large proportion of IgM^+ B cells has the capacity to recognize autoantigens, e.g. human IgG. In most people, two populations of IgM^+ precursor B cells with the capacitiy to produce rheumatoid factors can be distinguished in terms of their fine-specificity: one is multireactive, the other monospecific for human IgG-F_C-fragments. In healthy blood donors, the multireactive population dominates, whereas in patients with rheumatoid arthritis the ratio between the two frequencies is inverted and the monospecific B cell population is expanded. In patients with systemic lupus erythematosus, an additional IgG^+ population, binding dsDNA, appears, mainly during active phases of the disease. The observed changes in the B cell repertoire are characteristic for each disease.

Acknowledgement:

I would like to thank all colleagues, students and staff members who participated in the work reviewed here. The work was supported by DFG grants Pe 151/10 and Me 604/3.

References

1 Janeway CA, Travers P. The Immune system in health and disease. Oxford: Blackwell Scientific Publications, 1994.
2 Coutinho A. Immune or autoimmune? Clin Exp Immunol 1995;101(suppl.):3-5.
3 Cohen IR. The cognitive principle challenges clonal selection. Immunol Today 1992;13(11):441-4.
4 Cohen IR, Young DB. Autoimmunity, microbial immunity and the immunological homunculus. Immunol Today 1991;12(4):105-10.
5 Schwartz RS, Datta SK. Autoimmunity and autoimmune diseases. Chapt. 31 in: Paul WE, editor. Fundamental Immunology. 2nd ed. New York: Raven Press, 1989:819-66.
6 Naparstek Y, Plotz PH. The role of autoantibodies in autoimmune disease. Annu Rev Immunol 1993;11:79-104.
7 Parker DC. T cell-dependent B cell activation. Annu Rev Immunol 1993;11:331-60.
8 Bancherau J, Rousset F. Human lymphocytes: phenotype, proliferation and differentiation. Adv Immunol 1992;52:125-258.
9 Burnet FM. The clonal selection theory of acquired immunity. London: Cambridge University Press, 1959.
10 Theofilopoulos AN. The basis of autoimmunity: Part I. Mechanisms of aberrant self-recognition. Immunol Today 1995;16(2):90-8.
11 Cohn M, Langman RE. The protecton: the unit of humoral immunity selected by evolution. Immunol Rev 1990;115:11.
12 Jerne NK. Towards a network theory of the immune system. Ann Immunol 1974;125C:373.
13 Ziegner M, Steinhauser G, Berek C. Development of antibody diversity in

single germinal centers: selective expansion of high-affinity variants. Eur J Immunol 1994;24(10):2393-400.

14 Kuppers R, Zhao M, Hansmann M, Rajewsky K. Tracing B cell development in human germinal centres by molecular analysis of single cells picked from histological sections. EMBO J 1993;12(13):4955-67.

15 Schledz M, Hohlbaum A, Peter HH, et al. Specificity of monoclonal rheumatoid factors derived from patients with rheumatoid arthritis. [Abstract] Immunobiology 1992;186:93.

16 Menge T, Langenberg U, Peter HH, et al. Human monoclonal antibodies with specificities related to Rheumatoid Diseases. [Abstract] Immunobiology 1994;191:105.

17 Pascual V, Randen I, Thompson KM, Sioud M, Forre O, Natvig JB, Capra JD. The complete nucleotide sequences of the heavy chain variable regions of six monospecific rheumatoid factors derived from Epstein-Barr virus-transformed B cells isolated from the synovial tissue of patients with rheumatoid arthritis. J Clin Invest 1990;86:1320-8.

18 Casali P, Nakamura M, Ginsberg-Fellner F, Notkins AL. Frequency of B cells committed to the production of antibodies to insulin in newly diagnosed patients with insulin-dependent diabetes mellitus and generation of high affinity human monoclonal IgG to insulin. J Immunol 1990;144:3741-7.

19 Casali P, Prabhakar BS, Notkins AL. Characterization of multireactive autoantibodies and identification of Leu-1 + B lymphocytes as cells making autoantibodies binding multiple self and exogenous molecules. Int Rev Immunol 1988;3:17-45.

20 Mierau R, Gause A, Küppers R, Michels M, Mageed RA, Jefferis R, Genth E. A human monoclonal IgA rheumatoid factor using the VkIV light chain gene. Rheumatol Int 1992;12:23-31.

21 Lefkovits I, Waldmann H. Limiting dilution analysis of cells in the immune system. Cambridge: Cambridge University Press, 1979.

22 Wen L, Hanvanich M, Werner-Favre C, Brouwers N, Perrin LH, Zubler RH. Limiting dilution assay for human B cells based on their activation by mutant EL-4 thymoma cells: Total and anti-malaria responder B cell frequencies. Eur J Immunol 1987;17:887-92.

23 Schwab J, Lukowsky A, Volk H-D, Peter HH, Melchers I. Precursor frequencies for DNA-specific B lymphocytes in patients with systemic lupus erythematosus (SLE). Clin Exp Immunol 1994;96:450-7.

24 Rudolphi U, Hohlbaum A, Lang B, Peter HH, Melchers I. The B cell repertoire of patients with rheumatoid arthritis. Frequencies and specificities of peripheral blood B cells reacting with human IgG, human collagens, a mycobacterial heat shock protein and other antigens. Clin Exp Immunol 1993;92:404-11.

25 Melchers I, Peter HH, Eibel H. The T and B cell repertoire of patients with rheumatoid arthritis. Scand J Rheumatol 1994;24:153-62.

26 Rudolphi U, Lang B, Peter HH, et al. Frequency and specificity of B cells producing autoantibodies in the peripheral blood (PB), synovial fluid (SF) and synovial membrane (SM) of patients with rheumatoid arthritis (RA). [Abstract] Immunobiology 1991;183:188.

27 Kwekkeboom J, De Boer M, Tager JM, De Groot C. CD40 plays an essential role in the activation of human B cells by murine EL4B5 cells. Immunology 1993;79:439-44.
28 Winfield JB, Faiferman I, Koffler D. Avidity of anti-DNA antibodies in serum and IgG glomerular eluates from patients with systemic lupus erythematosus: Association of high avidity anti-native DNA antibody with glomerulo-nephritis. J Clin Invest 1977;59:90.
29 Pearson L, Lightfoot RW Jr. Correlation of DNA-anti-DNA association rates with clinical activity in systemic lupus erythematosus (SLE). J Immunol 1981;126:16-9.
30 Ehrenstein M, Longhurst C, Isenberg DA. Production and analysis of IgG monoclonal anti-DNA antibodies from systemic lupus erythematosus (SLE) patients. Clin Exp Immunol 1993;92:39-45.
31 Van Es JH, Gmelig-Meyling FHJ, Van de Akker WRM, Aanstoot H, Derksen RHWM, Logtenberg T. Somatic mutations in the variable regions of a human IgG anti-double-stranded DNA autoantibody suggest a role for antigen in the induction of systemic lupus erythematosus. J Exp Med 1991;173:461-70.
32 Winkler TH, Fehr H, Kalden JR. Analysis of immunoglobulin variable regions genes from human IgG anti-DNA hybridomas. Eur J Immunol 1992;22:1719.
33 Carson DA. Rheumatoid factor. Chapt. 11 in: Kelley WN, Harris ED, Ruddy S, Sledge CB, editors. Textbook of Rheumatology. 3rd ed. Philadelphia: WB Saunders, 1989:198-207.
34 Arnett FC, Edworthy SM, Bloch DA, McShane DJ, Fries JF, Cooper NS, Healey LA, Kaplan SR, Liang MH, Luthra HS. The American Rheumatism Association 1987 revised criteria for the classification of rheumatoid arthritis. Arthr Rheum 1988;31:315-24.
35 Roosnek E, Lanzavecchia A. Efficient and selective presentation of antigen-antibody complexes by rheumatoid factor B cells. J Exp Med 1991;173:487-9.
36 Borretzen M, Randen I, Zdarsky E, et al. Control of autoantibody affinity by selection against amino acid replacements in the complementarity-determining regions. Proc Natl Acad Sci USA 1994; 91(26):12917-21.
37 Pascal V, Viktor K, Randen I, et al. Nucleotide sequence analysis of rheumatoid factors and polyreactive antibodies derived from patients with rheumatoid arthritis reveals diverse use of VH and HL gene segments and extensive variability in CDR-3. Scand J Immunol 1994;40(1):125.
38 Saizawa KM, Melchers I, Eichmann K. Genetic control of B cell function. III. IgVH controlled polymorphism in the frequencies of B cells that recognize xenogeneic red blood cells. Eur J Immunol 1985;15(2):124-31.
39 Shin EK, Matsuda F, Ozaki S, Kumagai S, Olerup O, Ström H, Melchers I, Honjo T. Polymorphism of the human immunoglobulin variable region segment V1--4.1. Immunogenetics 1993;38:304-6.
40. Natvig JB, Randen I, Thompson KM, Forre O, Munthe E. The B cell system in the rheumatoid inflammation. New insights into the pathogenesis of rheumatoid arthritis using synovial B cell hybridoma clones. Springer Semin Immunopathol 1989;11:301-13.
41 Melchers I, Hummes R, Peter HH, et al. The B cell repertoire of patients with rheumatoid arthritis. [Abstract] Immunobiology 1993;189:121.

42 Rudolphi U, Rzepka R, Batsford S, Kaufman SHE, Von der Mark K, Peter HH, Melchers I. The B cell repertoire of patients with rheumatoid arthritis. II. Increased frequencies of IgG^+ and IgA^+ B cells specific for mycobacterial heat shock protein 60 or human collagen type II in synovial fluid and tissue. Arthr Rheum 1997 (submitted).

43 Klareskog L, Olsson T. Autoimmunity to collagen II and myelin basic protein: Comparative studies in humans and rodents. Immunol Rev 1990;118:285-310.

44 Cohen IR. Autoimmunity to chaperonins in the pathogenesis of arthritis and diabetes. Annu Rev Immunol 1991;9:567-89.

BERND ARNOLD, JUDITH ALFERINK, ANDREAS LIMMER AND GÜNTER J. HÄMMERLING

Self-tolerance

Introduction

The ability of the immune system to respond normally to pathogens but not to self-components is a central topic of immunological research. The parameters controlling the different consequences of antigen recognition by lymphocytes *in vivo*, activation or tolerance induction, must still be defined. A detailed knowledge of these processes is essential for therapeutic intervention in autoimmune diseases, organ transplantation, and in attempts to break non-responsiveness to tumor antigens. A new field of experimental approaches concerning self-tolerance and autoaggression has been opened by the possibility of generating transgenic and "knock-out" mice. In this review, mechanisms will be discussed which ensure the inactivation of self-reactive T lymphocytes in such mouse models.

Unlike immunoglobulins, T cell receptors (TCR) do not react with determinants on the intact three- dimensional protein structure but rather with peptides presented by major histocompatibility complex (MHC) class I or class II molecules.[1,2] Accordingly, "self" defined by T cells is probably represented only by that fraction of endogenous peptides which is bound with sufficient stability and presented above a certain threshold concentration by the particular MHC alleles of the individuum.[3] T cells directed against endogenous peptides binding only weakly to the respective MHC alleles and present at low abundance may not necessarily be tolerized.[4] These T cells cannot be activated under normal circumstances and are therefore not dangerous for the body. Indeed, self-reactive T cells are present in healthy animals. Autoimmune disease is only induced upon strong immunization of such T cells with the corresponding exogenous antigen or peptide.[5]

Since the antigen-specific receptors of lymphocytes are generated by random rearrangement of DNA segments, a sufficiently high number of specificities is created to react with any foreign but also of course with any self-structure. Thymic clonal deletion is the most important mechanism to eliminate those self-reactive T cells.[6] Encounter of antigen by thymocytes bearing the corresponding TCR leads to programmed cell death (apoptosis) *in situ*.[7] Any antigen, whether a naturally occurring self-protein or a viral protein introduced as a transgene product, causes deletion of the corresponding thymocytes if presented by bone marrow-derived cells above a certain threshold. Besides by clonal deletion, thymic tolerance can also be established by inactivation but survival of the thymocytes when the antigen is exclusively expressed on thymic epithelial cells.[8-11] In both cases, tolerance induction is based on the presence of the antigenic determinant during a stage of thymocyte maturation rather than a result of self-nonself discrimination by the individual T lymphocytes.

In addition to this central thymic tolerance, there is also a requirement for peripheral tolerance induction of T cells which have not been silenced in the thymus, possibly because the antigen was not present in the thymus. Since soluble peripheral antigens can be taken up by MHC class II positive cells in the thymus and induce thymic tolerance,[12,13] peripheral tolerance induction appears to be particularly important for $CD8^+$ cells, because soluble antigens usually do not enter the MHC class I pathway.[14]

Peripheral tolerance induction depending on various parameters

To learn whether peripheral tolerance induction occurs naturally and whether the underlying mechanisms could be useful for therapeutic purposes, transgenic mouse models were established by expressing a "foreign" protein under tissue-specific promoter elements in non-thymic cells.[15,16] During the study of these model systems it became clear that minute amounts of antigen expressed in the thymus are sufficient to induce tolerance. For example, K^b expression under the control of the rat insulin promoter (RIP-K^b)[17] or of the α-lactalbumin promoter[18] was undetectable in the thymus of the respective transgenic mice by histology and could only be seen on the mRNA level. In both cases, however, transfer of the transgenic thymi into thymectomized non-transgenic animals could induce tolerance to K^b in the respective chimeric mice.

Taken together, in order to establish a transgenic mouse system for peripheral T cell tolerance, one has to express an antigenic determinant above the threshold concentration for activation ensuring the necessity for tolerance induction and one has to prove that antigen expression is truly extrathymic.

On comparing the various published mouse models, we learned that the consequences of antigen expression in a tissue-specific manner are not yet predictable. Expression of different transgenes under the same rat insulin promoter sequences resulted in either tolerance, ignorance of the self-antigen, or induction of autoimmunity.[17,19-21] To gain further insight into the processes leading to peripheral tolerance induction, we have to consider the important parameters and have to develop model systems that allow us to vary the parameter(s) of interest and to keep others constant. The affinity of the TCR, the level of expression of the TCR and accessory molecules, and the developmental stage of the T cells encountering the antigen may be relevant in determining the consequences of contact between a T cell and a cell expressing the relevant antigenic determinant. In contrast to circulating T cells, recent thymic emigrants may be more susceptible to tolerance induction than to activation. This susceptibility could explain why early expression of antigen on pancreatic cells leads to tolerance, whereas late expression during development leads to autoaggression.[19] Properties of the antigen-bearing cell and the microenvironment of the cells are probably also of importance. The frequency, length and intensity of contact between T cells and the tolerizing tissue could be important and may depend on the number of antigen-bearing cells within the tissue and the accessibility of these cells to lymphocytes. In addition, the density of the self-antigen and the expression of adhesion molecules on the cell surface of a given cell type should contribute to the intensity of the contact. The two-signal concept of T cell activation has also to be taken into account. There is a qualitative difference in antigen presentation by macrophages and dendritic cells relative to non-lymphoid cells that determines whether activation or tolerance is the result of the contact.[22] Finally, the impact of lymphokines released into the microenvironment during inflammation should also be considered.

Based on these considerations, we have generated transgenic mice expressing the K^b antigen under control of various tissue-specific promoters such as the albumin promoter, the keratin IV promoter, or the GFAP (glial fibrillary acidic protein) promoter.[11,23-25] By immunohistology, cytofluorometry and mRNA analysis (determined by combined reverse transcriptase PCR), expression was restricted to extrathymic tissues: hepatocytes in Alb.K^b mice, keratinocytes of the skin and tongue in 2.4 kerIV.K^b mice, and in GFAP.K^b mice on cells of neuroectodermal origin such as glial cells and Schwann cells in the small intestine. All these mice expressed a K^b-specific TCR (DES-TCR) as a second transgene. Using TCR transgenic mice, one has the advantage of having many T cells expressing the same TCR which can be identified by anticlonotypic antibodies (Désirè-1). One should, however, not forget that when using TCR transgenic mice one can only follow the fate of T cells expressing one TCR specificity out of the entire repertoire. It is, therefore, essential to show that T cells bearing the transgenic TCR share the characteristics of the majority of T cells. DES-TCR^+ T cells are indeed able to reject positive skin and tumor grafts, cause graft-versus-host disease and develop into cytotoxic T cells *in vitro.*

All F1 mice with tissue-specific K^b expression crossed to Des-TCR mice were tolerant to K^b as indicated by their inability to reject K^b-positive skin grafts. Depending on the extrathymic tissue in which K^b was expressed, different mechanisms of tolerance could be observed.[24] In the 2.4 keratin IV.K^bxDes-TCR mice no phenotypic change with regard to Des-TCR and CD8 expression was found. The cells which were tolerant *in vivo* responded readily to K^b in a mixed lymphocyte reaction (MLR). In the GFAP.K^bxDes-TCR mice the Des-TCR and CD8 molecules were down-regulated. The remaining low level of Des-TCR was sufficient to allow activation *in vitro* in an MRL. In the Alb.K^bxDes-TCR mice the Des-TCR and CD8 molecules were also down-regulated but they did not respond to antigen *in vitro.* However, Des-TCR expression could be recovered by stimulation with cross-linked anti-CD2 antibodies.

So far, we could not distinguish between the contribution of the tissue on which the self-antigen is expressed and of the density of antigen on the individual cells to the observed manifestations of tolerance. Varying only the amount of antigen was possible by expressing K^b under the control of the C-reactive protein (CRP) promoter.[26] CRP is exclusively expressed in hepatocytes and can be strongly induced with LPS or IL-6.[27] In the absence of induction, K^b expression was hardly detectable on hepatocytes of CRP.K^b mice, but was clearly visible after induction with LPS. Despite the very low expression of K^b before induction the mice were tolerant as they failed to reject K^b-positive skin grafts. Thymic tolerance induction was excluded because CPR.K^b chimeric mice transplanted with a normal thymus were also tolerant in the periphery as documented by their acceptance of skin grafts.[26] Similar to the tolerance observed for Alb.K^bxDes-TCR or GFAP.K^bxDes-TCR mice,[11,23] this tolerance was accompanied by down-regulation of the Des-TCR on about 60% of the CD8 cells. Induction of the CRP promoter with LPS led to a strong increase of K^b expression on the hepatocytes. Within 3 days after induction, all CD8 cells had completely down-regulated their Des-TCR.[26] Since CD8 was also down-regulated, the CD2 marker had to be used for identification of T cells with down-regulated Des-TCR and CD8 molecules. In several mice the number of these CD2 positive cells was significantly decreased after induction with LPS, suggesting partial deletion of these cells. LPS by itself does not lead to down-regulation of the Des-TCR or to deletion as controlled by Alb.K^bxDes-TCR mice which have constitutive K^b expression on hepatocytes.

These results show that peripheral T cells can reach different levels of tolerance with regard to their phenotype and their capacity for reactivation. Anergy *in vivo* without phenotypic changes and with normal reactivity *in vitro* appears to be the weakest form of T cell tolerance, deletion of the autoreactive T cells the strongest one. These findings also demonstrate that tolerant T cells can still respond when they encounter increased amounts of the tolerogen.

Reversal of tolerance

Since the observed tolerance in our model systems was mainly based on non-deletional mechanisms, we wondered whether continuous contact with the tolerogen was required for maintenance of the tolerant state of these T cells. Tolerant T cells from Des-TCRx2.4ker.K^b mice were transferred into the antigen-free environment of lightly irradiated BALB/Cnu/nu mice. The recipients received K^b-positive skin grafts at various time points after cell transfer. Three days after transfer, most transferred cells were still tolerant as only 10% of the recipients rejected the graft. After 15 days, however, the parked cells were able to reject grafts in all recipients, demonstrating that persistence of antigen is required to keep the cells in their tolerant state.[28]

Since the tolerant cells could acquire responsiveness in the absence of antigen, we investigated whether tolerance could be broken *in vivo* without removal of the antigen. Methods for breaking tolerance *in vivo* could be helpful in situations where one would like to have a strong immune response and not tolerance induction, such as during growth of tumors. In order to see if molecules which help to activate virgin T cells can break tolerance, P815.K^b, K^b-transfected P815 (H-2^d) mastocytoma cells were co-transfected with the B7 or IL-2 gene. The P815.K^b.B7 transfectants grew in Des-TCRx2.4Ker-K^b tolerant mice like the parental P815.K^b tumor cells. However, the IL-2 producing tumor cells did not grow. In order to see if the rejection of the P815.K^b.IL-2 cells was due to activation of the tolerant K^b-reactive cells, the mice received either K^b skin grafts or P815.K^b tumor cells on their opposite flank 14 days later. All mice rejected the grafts. These studies demonstrate that simultaneous application of the tolerogen and IL-2 led to a systemic breaking of tolerance, resulting in activated Des-TCR positive cells with enhanced IL-2 receptor expression (own unpublished results).

Similar results were obtained with Des-TCRxAlb.K^b mice. Since these mice express K^b on hepatocytes, it was of interest to see whether activation of K^b-reactive cells could result in autoaggression against the K^b-positive liver. However, no infiltration of Des-TCR^+ T cells into the liver was found. These results have important implications for autoimmunity because they demonstrate that the presence of activated CD8 cells and the respective autoantigen do not necessarily result in autoaggression. It seems that additional factors are required, such as inflammation, damage of tissue, and/or cytokines.

Concluding remarks

The studies described here show that peripheral T lymphocytes can be rendered tolerant to antigens expressed exclusively outside the thymus. Depending on the amount of antigen and on the anatomical site of expression, T cells can reach various levels of tolerance which are distinct in their capacity of reactivation. Since tolerant T cells are able to regain their capacity to reject skin grafts in a neutral environment, we assume that multiple interactions with the antigen may be required to maintain a certain level of tolerance.

The finding that different tissues can induce different levels of tolerance could explain why some organs are more susceptible to autoimmune disesases than others. For example, potentially autoreactive T cells in a relatively mild stage of tolerance as described for the (Des-TCRx2.4KerIV-K^b) F1 mice are, of course, dangerous to an individual because tolerance may be easily broken under the influence of an invading pathogen.

Several important questions concerning peripheral tolerance induction remain. First, why is peripheral tolerance observed only in some experimental systems and not in others? One possible explanation for the different observations is the variation in the amount of antigen expressed. A quantitative relationship has been reported between the blood concentration of a self-antigen and the state of tolerance to its immunodominant, subdominant and cryptic peptides at the T cell level.[3]

Secondly, how do naive T cells meet the autoantigen on the nonlymphoid tissues? Direct interaction would require a rapid and complete circulation of all T cells. However, naive T cells in an adult individual mainly drain to the regional lymph nodes and have rarely been found to enter tissues. Since we have mostly discussed tolerance induction to peripheral antigens expressed constitutively during neonatal life, it has to be clarified whether early thymic emigrants can enter tissues and therefore differ in their migration behavior from naive adult T cells.

Thirdly, are recent thymic emigrants more susceptible than mature T cells which have been circulating for a longer time? These two types of T cells can indeed be distinguished phenotypically, suggesting a final extrathymic maturation step.[29] However, functional differences remain to be shown.

References

1 Babbit BP, Allen PM, Matsueda G, Haber E, Unanue ER. Binding of immunogenic peptides to Ia histocompatibility molecules. Nature 1995;317:359.

2 Björkman PJ, Saper MA, Samaraoui B, Bennett WS, Strominger JL, Wiley DC. Structure of the human class I histocompatibility complex antigen, HLA-A2. Nature 1987;329:506.

3 Cibotti R, Kanellopoulos JM, Cabaniols JP, Halle-Panenko O, Kosmatopulos K, Sercarz E, Kourilsky P. Tolerance to a self-protein involves its immunodominant but does not involve its subdominant determinants. Proc Natl Acad Sci USA 1992;89:416-20.
4 Gammon G, Sercarz E. How some T cells escape tolerance induction. Nature 1989:342:183-5.
5 Ben-Nun A, Wekerle H, Cohen IR. The rapid isolation of clonable antigen-specific T lymphocyte lines capable of mediating autoimmune encephalomyelitis. Eur J Immunol 1981;11:195-9.
6 Kisielow P, Boehmer H. Development and selection of T cells: Facts and puzzles. Adv Immunol 1995;58:87-210.
7 Murphy KM, Heimberger AB, Loh DY. Induction by antigen of intrathymic apoptosis of CD4+CD8+TCR10 thymocytes in vivo. Science 1990;250:1720.
8 Ramsdell F, Lantz T, Fowlkes BJ. A non-deletional mechanism of thymic self-tolerance. Science 1989;246:1038-41.
9 Salaün J, Bandeira A, Khazaal I, Calman F, Coltey M, Coutinho A, Le Douarin NM. Thymic epithelium tolerizes for histocompatibility antigens. Science 1990;247:1471.
10 Hoffmann MW, Allison J, Miller JFAP. Tolerance induction by thymic medullary epithelium. Proc Natl Acad Sci USA 1992;89:2526-30.
11 Schönrich G, Momburg F, Hämmerling GJ, Arnold B. Anergy induced by thymic medullary epithelium. Eur J Immunol 1992;22:1687-91.
12 Kyewski BA, Fathman CG, Kaplan HS. Intrathymic presentation of circulating non-major histocompatibility complex antigens. Nature 1984;308:196.
13 Zal T, Volkmann A, Stockinger B. Mechanisms of tolerance induction in major histocompatibility complex class II-restricted T cells specific for a blood-borne self-antigen. J Exp Med 1994;180:2089.
14 Braciale TJ, Morrison LA, Sweetser MT, Sambrook J, Gething MJ, Braciale VL. Antigen presentation pathways to class I and class II MHC-restricted T lymphocytes. Immunol Rev 1987;98:95.
15 Hämmerling GJ, Schönrich G, Momburg F, Auphan N, Malissen M, Malissen B, Schmitt-Verhulst AM, Arnold B. Non-deletional mechanisms of peripheral and central tolerance: studies with transgenic mice with tissue-specific expression of a foreign MHC class I antigen. Immunol Rev 1991;122:47-67.
16 Miller JFAP, Morahan G. Peripheral T cell tolerance. Ann Rev Immunol 1992;10:51-69.
17 Heath WR, Allison J, Hoffmann MW, Schönrich G, Hämmerling GJ, Arnold B, Miller JFAP. Autoimmune diabetes as a consequence of locally produced interleukin-2. Nature 1992;359:547-9.
18 Husbands SD, Schönrich G, Arnold B, Chandler PR, Simpson E, Philpott KL, Tomlinson P, Otreilly L, Cooke A, Mellor AL. Expression of major histocompatibility complex class I antigens at low levels in the thymus induces T cell tolerance via a non-deletional mechanism involving downmodulation of CD8 and T cell receptor. Eur J Immunol 1992;22:2655-61.
19 Adams TE, Alpert S, Hanahan D. Non-tolerance and autoantibodies to a transgenic self-antigen expressed in pancreatic b cells. Nature 1987;325:223-8.
20 Ohashi PS, Oehen S, Buerki K, Pircher H, Ohashi CT, Odermatt B, Malissen

B, Zinkernagel RM, Hengartner H. Ablation of "tolerance" and induction of diabetes by virus infection in viral antigen transgenic mice. Cell 1991;65:305-17.

21 Lo D, Freedman J, Hesse S, Palmiter RD, Brinster RL, Sherman LA. Peripheral tolerance to an islet cell-specific hemagglutinin transgene affects both $CD4^+$ and $CD8^+$ T cells. Eur J Immunol 1992;22:1013-22.

22 Schwartz RH. Costimulation of T lymphocytes: The role of CD28, CTLA-4, and B7/BB1 interleukin-2 production and immunotherapy. Cell 1992;71:1065-8.

23 Schönrich G, Kalinke U, Momburg F, Malissen M, Schmitt-Verhulst AM, Malissen B, Hämmerling GJ, Arnold B. Downregulation of T cell receptors on self-reactive T cells as a novel mechanism for extrathymic tolerance induction. Cell 1991;65:293-304.

24 Schönrich G, Momburg F, Malissen M, Schmitt-Verhulst AM, Malissen B, Hämmerling GJ, Arnold B. Distinct mechanisms of extrathymic T cell tolerance due to differential expression of self-antigen. Int Immunol 1992;4(5):581-90.

25 Arnold B, Schönrich G, Hämmerling GJ. Multiple levels of peripheral tolerance. Immunol Today 1993;14:12-4.

26 Ferber I, Schönrich G, Schenkel J, Mellor AL, Hämmerling GJ, Arnold B. Levels of peripheral T cell tolerance induced by different doses of tolerogen. Science 1994;263:674-6.

27 Ciliberto G, Arcone R, Wagner EF, Rüther U. Inducible and tissue-specific expression of human C-reactive protein in transgenic mice. EMBO J 1987;6(13):4017-22.

28 Alferink J, Schittek B, Schönrich G, Hämmerling GJ, Arnold B. Long life span of tolerant T cells and the role of antigen in maintenance of peripheral tolerance. Int Immunol 1995;7:331-6.

29 Hosseinzadeh H, Goldschneider I. Recent thymic emigrants in the rat express a unique antigenic phenotype and undergo post-thymic maturation in peripheral lymphoid tissues. J Immunol 1993;150 1670-9.

ANTONIO COUTINHO

Natural tolerance and self-assertion

Introduction

Over the last 20 years, powerful technologies have solved many of the questions pertaining to the molecular and cellular composition of the immune system. They have contributed relatively little, however, to our understanding of the selection of structural and functional lymphocyte repertoires, and therefore, to the basis of natural tolerance. Clearly, these are properties of whole organisms and not of single cells or molecules, and can thus be solved by no approaches other than those addressing the whole animal.

More recently, however, molecular genetics has also started providing new insights into these questions. Thus, both transgenic mice and mutants obtained by gene targetting in embryonic stem cells pose questions that the analytic empiricist can hardly avoid, even if primarily interested in the analysis of components. Rather than exclusively concerning the 'transgenic product' or the 'knocked-out function', the outcome of such experimental systems will necessarily pertain to levels of organization in the whole mouse. More importantly, as will become clear as we go along, these experimental systems cannot avoid 'going through' development and, therefore, cannot elude the most fundamental aspects of natural tolerance.

Many immunologists have long been convinced that the immune system is in continuous development throughout life and, consequently, that there is no particular reason to study its embryonic or perinatal characteristics, as opposed to the study of lymphocyte precursors in the adult. This is a general trend in the field, both in theory and in experiments, but it is incorrect on two accounts. First, if our interest is in systemic properties and not (only) in those of single cells, our observations should address the *organization* that is prevalent in embryonic development. Few would argue, indeed, that newborn and adult immune systems are comparable, regardless of the isolated characteristics of cells or gene products. Second, because there is evidence, partly discussed below, that embryonic and perinatal lymphocytes may be intrinsically different from those produced in the adult (as also happens, for example, in the erythroid pathway of differentiation of hemopoietic cells). In turn, these two aspects are intimately related, since global properties emerge from the organization of components with characteristic properties and 'local rules'. They must be kept well separated, however, to avoid confusion in the levels of description of the phenomenology we address.

Here, I shall argue that natural immunological tolerance is based on assertion of the embryonic self, mediated by mechanisms that retain a memory of the molecular composition of the organism during development. These mechanisms dominantly select V-region repertoires and effector functions of autoreactive cells that are not aggressive to the target structures, and are recruited into a dynamics of equilibrium, thus ensuring 'dominant tolerance'. I shall also attempt to delineate a model for structural and functional repertoire development and cell interactions that accounts for some current observations and may coherently explain the physiopathology of autoimmunity. Some implications of these views will also be discussed.

Tolerance is immune activity in equilibrium

The vertebrate immune system has evolved under many constraints and pressures. Anti-infectious defence has certainly been one, but it is often forgotten that we live in perfect symbiotic equilibrium with a large number of microorganisms. The number of bacteria colonizing the surfaces of our body (skin and mucosae) is estimated to be one order of magnitude higher than the number of (eukaryotic) cells in all our organs and tissues. Obviously, the very effective antibacterial mechanisms in vertebrates – including specific immunity as well as 'nonspecific' effector molecules and cells already fully developed in invertebrates – do not exclude those symbiotic bacteria from the body, and it would thus be appropriate to speak of 'tolerance'. Normal individuals, however, would not be considered 'tolerant' to those microorganisms in the conventional sense, that is, unable to mount a specific immune response. Thus, either in the mucosa-associated lymphoid tissues of normal individuals, or even systemically, many immune activities can be recorded that actually represent specific 'responses' directed at the respective bacterial antigens. Furthermore, the unfortunate consequence of immune deficiencies (as seen in AIDS patients), with the loss of such 'responses' to the symbiotic flora, is that such 'normal' bacteria often lead to severe damage and death. In other words, given that no response to the symbiotic flora leads to disease and death, normal individuals must continuously keep responding, although in manners that do not result in the elimination of such bacteria from the body. For the definition of tolerance that I shall adopt here, normal individuals are indeed tolerant to symbiotic bacteria, for they coexist at equilibrium, in spite of the ongoing immune responses provoked by the bacteria which are actually necessary to maintain such an equilibrium.

This notion can be readily applied to natural tolerance to body tissues: here also, the existence of specific (auto)immune activities is not at all contradictory to the coexistence with a normal body composition and functions, in physiological equilibrium. This can be understood on several accounts. First, many kinds of activities of lymphocytes and their products are not destructive of their molecular or cellular targets and are, thus, not aggressive. Not only the group of cytokines defined as 'hemopoietic growth factors', but also a number of T cell-derived lymphokines, often also produced by other cell types, have been known for some time to enhance survival and performance, even proliferation, of the respective target cells (*e.g.*, EGF, KGF, TGF-β1, etc.). We know today that some of the supposedly most 'aggressive' T cell-derived lymphokines (*e.g.*, IL-1, IL-6, TNFα, etc.) actually participate in the normal regulation of tissue development and organization, from very early stages in development. Furthermore, most such effector molecules of activated lymphocytes are currently designated as 'immunoregulatory' (*e.g.*, TGF-β1) for they mediate both activating and suppressive actions in immune responses.

Most immunologists, however, continue to view all immune activities as necessarily aggressive and destructive, in spite of repeated demonstrations of (1) antibodies that are 'facilitating' the survival of the respective target structures, such as tissue grafts; (2) antibodies that enhance infectivity of phages and parasites; (3) the whole world of 'suppression', where immune responses are indeed inhibitory or destructive, but of other (target-aggressive) immune responses; (4) the recently rediscovered notion of 'immune deviation', where T lymphocyte responses may well lead to consolidation, rather than elimination, of a parasitic infection, as seems to be the general rule with 'Th2'–type responses for intracellular parasites. Perhaps this is the curse laid upon the discipline by its birth as a branch of microbiology. Once again, only by assessing the outcome of immune activities *in vivo* (elimination *versus* persistence of target structures) can we conclude as to tolerance or immunity, and clearly not by *in vitro* tests of isolated components (that is, facilitating and 'rejecting' antibodies may bind equally well to the target tissues, 'Th2' and 'Th1' cells may proliferate equally well *in vitro* in response to the appropriate antigens).

With reference to the physiopathology of autoimmunity, the question has an interesting twist. Thus, either type of immune activity – destructive or stimulatory – can actually cause disease, as is the case with antithyroid cell antibodies which can either stimulate or inhibit the function of the organ. An antireceptor antibody that overstimulates thyroid function, however, can hardly be considered destructive of the target organ, and yet it causes a serious autoimmune condition. A radical position could actually maintain that antibody molecules (with the exception of those that display enzymatic activities) are never destructive or protective *per se*, but only in the context and with the participation of other (phagocytic) cells, or of nonspecific (enzymatic) systems, like complement. Hence, 'class regulation' of lymphocyte effector functions should again be of central importance in the understanding of natural tolerance. On the other hand, the same argument would suggest that the concentration of autoantibodies (particularly if agonists of cellular receptors, or if neutralizing biologically active mediators) must be kept under stringent control (by other antibodies?), regardless of their class. The corresponding ideas could be developed for many of the effector activities of T lymphocytes.

The idea of equilibrium also implies that, while both partners of the interaction stimulate each other, the products of their activities are mutually neutralizing (or destructive), so that the characteristic dynamics for that particular equilibrium are maintained. That is, even if some effector immune activities are indeed destructive of their target structures, the regenerating ability of the latter may well compensate for the former, which are in any case the result of stimulation by the latter. The partners are mutually dependent! This is important to consider, both in respect to equilibria with replicating organisms, as well as in the case of tolerance to self components that are continuously synthesized *de novo* and turned over.

Self tolerance is a developmental property

Burnet's solution to self/nonself discrimination was simple and coherent, for it required only two premises and one postulate, all of which were entirely acceptable at the time: (1) all lymphocyte specificities are exclusively produced during embryonic life; (2) all self antigens, but no nonself ligands, are already present in the embryo. Hence, self/nonself discrimination can readily be ensured by postulating a developmentally regulated switch in the mode of lymphocyte responses to antigen recognition: inactivation in embryonic life versus activation in the adult. Many observations have since confirmed the correct assumption that the establishment of tolerance is essentially limited to development, but it also became clear that Burnet's first premise is wrong. In mammals at least, new lymphocytes and specific reactivities are produced from uncommitted precursors throughout life.

This apparently small deviation from the original premises invalidates Burnet's proposal, because the time window for generating new lymphocyte reactivities cannot be reduced to the period of exclusive exposure to self antigens. Thus, all lymphocytes produced after birth will be equally exposed to self and to the foreign antigens present in the internal environment and against which normal animals respond. It follows that *each individual cell cannot be competent to ensure self/nonself discrimination because the difference between self and nonself is historical and antedates the existence of the very lymphocytes that must carry on the discrimination.* In other words, *what lymphocytes must do is not to discriminate between antigens* that are present and those that are absent from the organism, *at the time the lymphocytes are produced and reach the periphery*, since newly formed lymphocytes will encounter both self and nonself. *Newly formed lymphocytes must instead discriminate between antigens that were present during development and those that were not, long before the lymphocytes themselves were ever produced.* To ask individual cells to perform at these levels, as currently done by the prevailing 'recessive' hypothesis on self-tolerance, brings immunology into the realm of miracles and religion.

In Burnet's proposal, most lymphocytes in a normal individual – which, as we now know, are produced in post-embryonic life – would be activated by self antigens, just as if they were directed at nonself. This difficulty has been dealt with in several ways. First, by Lederberg, who took Burnet's postulate of a developmental switch in the mode of lymphocyte responses and applied it to single cells rather than to the whole organism. This is a serious mistake that has been (and continues to be) at the origin of much confusion. Thus, according to this view, *there is no apparent reason why the establishment of self tolerance should be restricted to embryonic life*: as lymphocytes continue to be produced throughout life, it should be roughly as easy (or as difficult) to tolerize adult or newborn animals. We know this is wrong, and Lederberg himself explicitly acknowledges that *only 'immature' immune systems are tolerized*, in spite of his postulate that *only 'immature' cells are tolerizable.* An alternative idea, which is often confused to some extent with the previous one, is that differentiating lymphocytes in the adult are secluded from all external antigens. In this case, the primary lymphoid organs are equivalent to the embryo, and all cells can be screened for autoreactivity and inactivated 'à la Lederberg'. In these views, the problem is not so much how to keep external antigens out of the thymus and bone marrow, but rather, how to get all self antigens into those organs. The recent work with transgenic mice, expressing 'neoantigens' in organ-specific ways, has come to confirm our prejudices that such models are also wrong, by demonstrating that natural tolerance most often concerns tissue-specific antigens that are never present in the thymus (or bone marrow).

One of the first transgenic systems in which a neoantigen (in this case, a viral protein) was expressed under the control of a tissue-specific promoter (the rat insulin promoter, which directs expression to insulin-producing pancreatic β-cells) has also provided good experimental evidence to confirm the basic notion that natural tolerance is exclusively acquired in embryonic and perinatal life. Thus, Hannahan and colleagues have shown that transgenic lines in which the transgene was expressed before birth did become naturally tolerant, while those which expressed the same transgene after birth did not and went on to rapidly destroy the 'target' cells.

From these considerations, it becomes quite clear that *we will have to consider in some detail the developmental aspects of the immune system, for this is the uniquely critical period when self is 'learned'*. The hope is that the characteristics of the system during the learning period will shed light on the mechanisms underlying the establishment of tolerance and, possibly, on those maintaining it in the adult as well. There is a set of current views as to the basis of natural tolerance which are not concerned with developmental aspects (and must therefore be wrong, according to my argumentation). This is a heterogeneous group of notions that ascribe to the differential 'professionalism' of 'antigen-presenting cells' the basic decision between induction *versus* paralysis (anergy, in the most popular nomenclature). These models are also uniformly based on the inactivation of autoreactive cells, and can thus be discussed together with those that are explicitly based on 'recessive' mechanisms. In short, there are only two sides when considering and discussing natural tolerance: one is based on the absence and/or functional competence of single (autoreactive) clones and must thus be 'recessive'; the other invokes 'dominant' mechanisms of regulation, and it requires the availability and activation of autoreactive lymphocytes, that is, it is based on 'physiological autoimmunity'. In short, and whatever their details, the 'recessive' hypotheses are 'clonal' while the 'dominant' models are 'supraclonal' and argue that the problem will never be solved by restricting our approaches to clonal analyses.

Before discussing the models of natural tolerance, let us review a few basic principles:

1. Normal mammals produce new lymphocytes throughout life; these display 'unpredicted' reactivities, and a particular frequency of autoreactivities.
2. Lymphocytes differentiate to maturity and antigen-reactivity in primary lymphoid organs where they are not exposed to most tissue-specific self antigens. These newly formed lymphocytes, emigrating from the primary organs to the periphery, will thus be exposed, as naive immunocompetent cells, to both autologous antigens (towards which tolerance must be maintained) and nonself antigens that are present in the organism (against which responses should be ensured). It follows that single, newly formed lymphocytes have no way of telling the difference.
3. Mechanisms must exist throughout life which can educate newly formed autoreactive lymphocytes to be non-aggressive towards antigenic structures that were present during development, anywhere in the body (*e.g.*, inside the pancreas, as seen above). These mechanisms must display antigenic specificity and thus embody

what could be designated as a 'memory of embryonic self'. Long-term memory of the embryonic self, however, does not necessarily imply that the supportive cells are as long-lived as the memory itself, for memory can be maintained by a dynamic process of recruiting, for similar non-aggressive functions, new lymphocytes of the appropriate specificity, while letting those that previously ensured the process decay.

4. Embryonic and adult immune systems must differ in the initial conditions under which self is first positively identified; these conditions must impose a non-aggressive mode of reaction towards the set of antigens that are then recognized, and be compatible with the mechanisms that maintain the dominance of such mode of reaction throughout adult life, so that newly formed lymphocytes can be recruited into similar functions and, therefore, dominantly maintain self-tolerance.
5. Dominant mechanisms that ensure self-tolerance towards the set of antigens first recognized in the embryo must not interfere with the immune responses to antigens encountered later in life.

A model of developmentally acquired dominant T cell tolerance towards self-antigens

Robust mechanisms, such as those required to establish self-tolerance, are likely to be evolutionarily derived, transmitted in the germ-line and read-out as developmentally-related programs for gene expression. T cell specificity can be developmentally 'programmed' by ordered rearrangement of TCR gene segments, and other mechanisms ensuring the precise structural composition of T cell specificities at early developmental times (germ-line segmental homologies that favor precise rearrangements, developmental controlled expression of TdT, etc.). There is ample evidence for such 'programs' in connection with $\gamma\delta$TCRs–the first to be produced in the embryo, but only one case is known for $\alpha\beta$TCRs. T cell function, on the other hand, can be directly 'programmed' by developmentally regulated cytokine gene expression. Furthermore, given that T cell effector functions can be determined by the microenvironmental conditions under which cell activation takes place, another strategy to direct T cell function in development would be to 'program' the cellular composition (and thus, the microenvironment) of the thymus where T cells first have the opportunity to see antigens and where the fate of individual T cells is determined. The present model uses all these alternative strategies.

The first lymphocytes to be formed in the embryonic thymus are $\gamma\delta$T cells with anti-self specificity. These are known to differentiate under very precise time schedules, and later migrate to the skin, where they show the ability to produce growth factors for epithelial cells. It follows that germ-line conditions ensure autoreactivity of the very first T cells with antigen-presenting cells in the thymus, themselves epithelial in nature. It is also well known that the development of the thymic stroma depends upon productive interactions with differentiating lymphocytes. Furthermore, it has been shown that thymic epithelial cells are activated to produce mediators, such as TGFβ-1, by epidermal growth factor (EGF). The model postulates that the first $\gamma\delta$T cells complete their differentiation upon recognition of self-antigens presented by TE cells, and become activated to produce EGF. In turn, EGF activates TE cells to further differentiate and produce TGFβ-1. All these processes, including EGF expression by the original $\gamma\delta$T cells, are 'programmed' in the germ-line, to the extent that no diversity is allowed in the $\gamma\delta$TCRs involved, in which no polymorphism has been noted to date, even in rather distant mouse subspecies. It should be expected that the respective ligands are also nonpolymorphic and expressed by the TE.

The conditions have now been created to let other T cells ($\alpha\beta$) appear and differentiate. With the exception of the Vα14Jα218 TCR-chain, there are no examples of developmentally 'programmed' $\alpha\beta$TCR specificities. The model would be more robust if these would exist (to ensure a high frequency of autoreactivity in the earliest $\alpha\beta$T cells), but it does not strictly require them, relying on random diversity to ensure self-reactivities. Moreover, absence of TdT expression results in a perinatal 'promiscuous' T cell repertoire that has been shown to be particularly rich in autoreactivities. Through such control mechanisms, precursors of $\alpha\beta$T cells that recognize self-antigens expressed on TE cells with high affinity appear at high frequency. Given that TE cells are excellent APCs, such T cells are activated. Given that TE APC have previously been activated to produce TGF-β1, autoreactive T cells activated in the embryonic thymic environment are activated towards non-aggressive effector functions (production of TGF-β1, perhaps IL-4 and other non-inflammatory cytokines).

As amply demonstrated, hemopoietic APCs (dendritic cells and macrophages/monocytes) are very competent in deleting differentiating T cells in the thymic medulla. It has also been repeatedly demonstrated, however, that most intrathymic T cell deletions do not occur until days or weeks after birth in the mouse. The model postulates, therefore, that the embryonic and neonatal thymus is deficient in the compartment of hemopoietic APCs, and thus unable to delete autoreactive T cells produced (and actually activated) in that period. Such T cells, already committed to effector functions because of their intrathymic activation, thus exit the thymus and reach the periphery, where they will encounter (at least a subset of) the same self-ligands towards which they can react and which activated them intra-thymically. Hence, these first αβT cells that reach the periphery are committed to non-inflammatory functions and locally reactivated in the tissues. These cells, however, do not cover the universe of tissue-specific antigens, for they were exclusively selected on the basis of antigens expressed by the TE and, thus, can only react in the periphery with ubiquitous self-antigens.

T cell production, however, continues in the thymus, giving rise through random diversity to T cells directed at foreign antigens and to T cells directed at tissue-specific self-antigens that are absent from the thymus. By conventional mechanisms of positive selection (which only require TE but do not involve affinities high enough to activate the cells – in which case they fall into the above category), both anti-foreign and anti-self (tissue-specific) T lymphocytes complete differentiation, reach the resting state, and exit the thymus as functionally uncommitted, newly formed resting immunocompetent cells. In the periphery, T cells with specificity towards antigens that are absent from the organism (foreign) will remain resting and will eventually decay. In contrast, tissue-specific T cells will encounter self-antigen and will be activated, but this process will take place in the presence of ubiquitous peptide-specific T cells that are locally (re)activated by antigens presented on the same tissue cell. Following the current principles of feed-forward loops determining cytokine production, the model postulates that tissue-specific self-reactive T cells are thus recruited into similar effector functions as those that had been imposed in the thymus on the first autoreactive T cells. In other words, the developmental principle here is the 'transfer' to the periphery of the 'local microenvironmental' conditions that characterize the embryonic thymus and are responsible for the activation of non-aggressive functions. This transfer, however, is selectively done onto tissue-specific self-reactive T cells that have recently emigrated from the thymus. This process ensures the perinatal build-up of a repertoire of 'regulatory' tissue-specific T cells that can dominantly maintain tolerance throughout life by the continuous recruitment of other recent thymic emigrants with similar specificity. Recent findings in sheep have demonstrated a unique pattern of T cell recirculation in the embryo, leading to a very abundant presence of T cells in nonlymphoid tissues, in sharp contrast to the adult. This adds one more developmentally related feature of the embryonic/perinatal immune system that seems to contribute to the mechanisms establishing natural tolerance.

Several developmental changes contribute to the shift from 'tolerance to immunity' that typically occurs during the first few weeks of life in the mouse. First, the compartment of thymic hemopoietic APCs becomes mature and competent, so that autoreactive T cells that recognize antigens expressed on the APCs no longer leave the thymus. Ubiquitous antigen-specific self-reactive regulatory T cells will no longer be available in the periphery. Tissue-specific T cells continue to exit the thymus, but these will be 'educated' into regulatory functions in the periphery, by the set of T cells previously selected, and actually contribute to the maintenance of a 'dynamic memory of self'. Second, the antigen-dependent peripheral recruitment of newly produced T cells into effector functions that mimic those of the pre-existing cells present in the same micro-environment can only take place within a finite time period after exiting the thymus. If T cells do not encounter specific antigen soon after reaching the periphery, they will no longer be recruitable into regulatory functions. As the numbers of T cells cease to increase, when the animal stops growing, fewer and fewer peripheral T cells will thus be in the 'recruitable' time window. It follows that the dynamics of T cell acumulation in the periphery, a developmental process 'par excellence', is indirectly responsible for the shift from tolerance to immunity.

Since hemopoietic APC delete intrathymic T cells that specifically interact with their surfaces, this particular tissue represents the only one in the body for which no 'regulatory' cells exist. They constitute, therefore, the best presenting cells in the induction of immune responses to foreign antigens that will accordingly be of a predominantly inflammatory type.

A model for B cell tolerance

In contrast to T cells, B lymphocytes display monotonous effector functions. The distinction between tolerance and immunity must thus be ascribed to other characteristics. On the other hand, B cells, but apparently not T lymphocytes, undergo somatic mutation and affinity maturation. One fundamental difference between antibodies to foreign autoantigens concerns precisely somatic mutation. Normal individuals produce large amounts of autoantibodies, but they are all encoded by germ-line, unmutated VH- and VL-genes. In contrast, antibodies produced during immune responses to foreign antigens (and in the case of antibody-mediated autoimmune diseases) accumulate somatic mutations and increased affinities. The problem of B cell tolerance, therefore, is reduced to the question of how to avoid somatic mutation and affinity maturation in autoantibodies. As repeatedly shown in normal individuals, this control is perfectly ensured, even when the autoantibody responses are induced by conventional, linked and restricted T-B cell collaboration.

There are at least two alternative mechanisms to avoid affinity maturation of autoreactive B cell clones. Let us consider them one at a time. A simple manner to avoid affinity maturation following somatic mutation is to limit the extent of clonal expansion. Thus, most mutations will result in 'loss of function', and therefore, many mutational events must accumulate before increased affinities can be scored and selected. The problem can, therefore, be translated into how to limit clonal expansions of autoreactive B cell clones, while allowing for large amplifications in the responses to foreign antigens. If we search for possible indications in the known developmentally regulated mechanisms determining the expression of the perinatal B cell repertoire, the following striking properties of these repertoires have been described. First, there is an extraordinarily abundant developmentally controlled overexpression of the most D-proximal V-H-gene (VH7183.1). This is determined by two convergent mechanisms: on the one hand, there is a very marked preference of rearrangement of that VH-gene, and on the other hand, B cells expressing antibody receptors bearing this VH-region undergo marked positive selection (survival and/or expansion) during embryonic life. Secondly, perinatal antibodies are characterized by CDR3s that are devoid of N-region additions, and often show little or no trimming at the ends of the recombined gene segments. In other words, there are developmental genetic programs to ensure 'germ-line' expression of antibodies bearing VH7183.1 Heavy-chains. Functional analysis of these antibodies has revealed their extraordinary ability to react with other immunoglobulins, that is, the selectively over-expressed 'germ-line antibodies' are highly 'connected', a property that had previously been described for the whole perinatal repertoire. It does not require much imagination to realize that a highly connected set of antibodies and cell receptors is not compatible with isolated clonal expansions.

As long as the overall levels of connectivity in the system remain high, all V-regions (reactive with self- or foreign antigens) are likely to be treated similarly. However, as soon as those levels fall below some critical threshold, autoreactive B cell clones that find 'internal ligands' will certainly have an advantage to remain in the original network organization. This is a second (dynamic) property of the germ-line encoded network organization, namely, that autoreactive B cells produced in this period are recruited into the neonatal network, with exclusion of (resting) B cells which recognize foreign antigens. As proposed before and demonstrated in principle by computer simulations of the appropriate models, the post-natal fall in germ-line connectivity is a necessary consequence of network dynamics per se, which actually determine the size and connectivity of the steady-state (adult) network. Thus, since B cell responses follow bell-shaped curves for receptor occupancy, the original connectivity raises the levels of circulating 'connected' antibodies to supraoptimal levels, so that inactivation of the most connected clones is automatically initiated. As the overall connectivity decreases below levels that maintain all clones in the network, autoreactive B cells will be selectively recruited. Thus, newly produced autoreactive B cells are activated and secrete antibodies at a high rate; these autoantibody clonotypes, circulating in the serum, will be in selective condition, quantitatively and qualitatively, to engage in productive interactions with complementary clones, thus incorporating the network structure. For purely dynamic reasons, therefore, the neonatal network will mature to be shaped around the significant autoreactivities, a set of self-antigens and respective autoantibodies that we have designated 'antibody homunculus', following the designation introduced by I. Cohen for autoreactive T cell repertoires. Interestingly, this epigenetic organization and dynamics exclude from the network B cell clones that remain resting because they did not find productive complementarities inside the body; that is, B cells directed at foreign antigens are resting cells, disconnected from a network organization.

These conditions in the mature, adult system, derived from the recognition of the set of autoantigens present at a critical period in development, would obviously result in the inability of autoreactive B cells to engage in significant clonal expansions, while allowing for unbound amplifications of B cell clones directed at non-self antigens. In other words, acquisition of tolerance is developmentally restricted but is dominantly maintained in the adult by supraclonal mechanisms.

Interestingly, another putative property could be derived from the characteristic neonatal network organization, namely, the 'homeostatic' control of the total number of B cells and of the circulating concentrations of immunoglobulins. In turn, these 'natural antibodies' should be expected to display autoreactivity, and high levels of degeneracy or 'multireactivity', just as has been found experimentally. The characteristics of these antibodies would make them good candidates to engage in second-order regulatory interactions with T cells and respective receptor V-regions, as previously described empirically and recently made explicit in a formal model.

The second general type of mechanisms limiting autoreactive B cell clonal expansions in the adult could be derived from the characteristics of a tolerant T cell compartment, as discussed above. Thus, B cells reacting with autoantigens would present the respective peptides to T cell recognition. It is clear from the model above that such autoreactive T cells would display characteristic functional phenotypes, designated as 'regulatory'. It could thus be hypothesized that such T cells, or a subset of them, display a type of helper activities towards B cells that activate the target cell to terminal differentiation and antibody secretion while suppressing clonal expansion. Again, this type of effector function would read as 'dominant', and would be entirely compatible with the predominance of autoantibodies among circulating immunoglobulins of normal individuals.

These two alternative modes of controlling B cell clonal expansion would suggest different pathogenetic mechanisms in antibody-dependent autoimmune disease, which is accompanied by very large clonal expansions, allowing for extensive somatic mutation and significant affinity maturation. For a pure network model, autoimmune disease would often result from defects in antibody connectivity, be these localized – as in organ-specific autoimmunity, or generalized – as in systemic autoimmune diseases. For the model based on T cell control, autoimmune disease would instead result from defects in 'regulatory' T cell repertoires or function. The therapeutic effects of normal immunoglobulins at high doses in a variety of autoimmune conditions could be used in support of the former, while the fact that all autoimmune diseases are T cell dependent is perhaps more compatible with the latter. It is not unlikely, however, that both are right, if the establishment and maintenance of physiological levels of connectivity are, in part at least, T cell dependent, if such helper cells are simultaneously 'regulatory', and if natural antibodies mediate significant interactions with T cell receptors so as to participate in the selection of their repertoires and of the functional phenotypes of the respective cells.

Acknowledgements

The work reported here was carried out with support from the EC, AFIRST, EFG, and Sidaction. I thank all my colleagues for stimulating discussions in the development of these models.

G. COAKLEY, M.A. HALL AND J.S. LANCHBURY

Perspectives on HLA and disease

Autoimmune disorders are common world-wide, and are major causes of severe morbidity and premature mortality. There is clear evidence for an important genetic component, and great efforts have gone into the investigation of human leukocyte antigen (HLA) associations over the last twenty years. Our understanding of these diseases has improved enormously as a consequence. Several original hypotheses of disease causation have arisen from this work, some of which have now been tested. This review will concentrate on recent advances in molecular biology and immunology which have provided a framework for understanding the role of the major histocompatiblity complex (MHC) in the development of rheumatoid arthritis, insulin dependent diabetes and coeliac disease.

Nature of Polymorphism

LEVELS OF POLYMORPHISM OF CLASS I AND II The human MHC maps to the short arm of chromosome 6 in the region designated 6p21.3, which represents the most intensively studied area of the human genome. This region occupies approximately four megabases of DNA and contains at least 100 genes.[1] Previously, the MHC has been operationally divided into the class I, class II and class III regions which encode the classical transplantation antigens, the immune response gene products and a diverse grouping of genes including several complement components, respectively. The location of genes of similar function to particular areas of the MHC is probably a legacy of gene duplication during evolution. Although still useful, this distinction is breaking down as more genes with novel functions are described, such as the peptide transporter genes which map to the HLA class II region but are apparently involved in delivering peptides to HLA class I molecules.

This review is primarily concerned with the HLA class I (A, B, and C) and class II (DR, DQ, and DP) genes. Each of the class II regions DR, DQ and DP contains at least two genes coding for α and β chains, denoted A and B, respectively. The function of MHC molecules is to collect peptide fragments inside the cell and transport them to the cell surface, where the peptide-MHC complex is presented to T cells. The DMA and DMB genes which map to the class II region are related structurally to both class I and II but are apparently involved in the loading of other HLA class II molecules with peptides.[2]

Immunological interest in the MHC stems from its role in controlling the specificity of immune responses to protein antigens. Many HLA class I and class II genes and their products are highly polymorphic, with loci such as HLA DRB1 currently associated with 77 alleles, and HLA B with 85.[3] At most of the polymorphic HLA loci, the heterozygosity is about 90% and most of the alleles occur at a frequency below 0.16. The extent of polymorphism has proved useful in uncovering relationships between specific alleles and predisposition to disease, but this process is hampered by linkage disequilibrium. This is a phenomenon by which particular combinations of alleles tend to be inherited as blocks known as haplotypes. Family studies have shown that recombination in the HLA region is rare, and thus a complete set of alleles is usually inherited as a haplotype. This often leads to difficulties in identifying the precise locus of susceptibility within a linkage group of positively associated alleles. One solution is to examine these disease associations in a variety of ethnic groups where the haplotypes carrying susceptibility alleles may be distinct.

EXTENSIVE AMINO ACID POLYMORPHISM IN THE ANTIGEN BINDING GROOVE As outlined above, the HLA region is extremely polymorphic, but this polymorphism is not evenly distributed throughout the expressed molecule, and is seen to the greatest extent in the antigen binding groove. The expressed HLA class I molecule comprises a heavy chain encoded by the class I genes of the HLA region and a light chain encoded by the β2 microglobulin gene on chromosome 15. The heavy chain is divided into α1, α2 and α3 domains. The surface of the molecule that binds peptide faces away from the cell, towards the receptor of the incoming 'cytotoxic' T cell. The groove, making up the antigen recognition site, is bordered by two α helices , and the floor is a β pleated sheet.[4]

The class II molecule is similar in conformation to the class I molecule, although their respective genes are only distantly related. The antigens (DR, DQ and DP) are composed of an α and a β chain, each of around 200 amino acids, and coded for by A and B genes, respectively. The extracellular part of each chain is divided into two domains, α1, α2 and β1, β2. These combine to form an αβ heterodimer, whose antigen recognition site binds and presents antigen to the 'helper' T cell. In a similar manner to class I, the crystal structure of HLA-DR1 has confirmed that the genetic variation of HLA class II alleles is particularly pronounced at the amino acid residues of this antigen recognition site.[5] Presumably, this variation has arisen because of the selective advantage gained from the ability to bind a wider range of antigens, and in turn present them to T cells. Mutation analysis supports this view with a strong bias towards replacement amino acid substitutions in the antigen binding site, suggesting that natural selection is responsible for the polymorphism observed at class I and class II loci.[6,7]

HLA associations with immune mediated disease

Many diseases have now been shown to have an HLA association, but this review is primarily concerned with rheumatoid arthritis (RA), insulin dependent diabetes (IDDM) and coeliac disease (CD), which may be regarded as prototypes.

RHEUMATOID ARTHRITIS The overall genetic contribution to the development of RA is estimated to be around 30%. Of this, probably only around half is accounted for by genes in the HLA region. A genome screening approach in multicase families is currently being undertaken in Oxford, and has to date revealed contributions to genetic susceptibility arising in several chromosomes, although none more important than the MHC (Paul Wordsworth, personal communication). It is clear from serological, cellular and molecular studies that the greater part of the HLA association is accounted for by molecular polymorphisms which map to the HLA class II region, and to the HLA-DRB1 locus in particular. Positive associations of HLA class I alleles with RA have been reported, but it is likely that the majority (such as B44, B60 and B62) are accounted for by linkage disequilibrium with HLA-DRB1. It remains a possibility that other MHC-encoded loci may modify the disease course, and evidence for this will be discussed.

RA in hospital clinics in Northern European, Chinese, New Zealand Polynesian and most US populations is associated with DRB1*0401, 0404, 0405 and 0408, but is not associated with the closely related subtypes DRB1*0402, 0403 and 0407.[8-10] Of the non-DRB1*04 alleles, DRB1*0101, 1001 and 1402 are associated with RA. Gregersen[11] outlined the 'shared epitope' hypothesis in 1987, proposing that the unit of HLA association with RA was a third hypervariable region (HVR3) pentapeptide motif which would be shared in a conserved form between RA associated alleles. The DRB1*0101 and 1402 alleles share the QRRAA HVR3 motif at positions 70 to 74 with DRB1*0404, 0405, and 0408. DRB1*1001 carries a related sequence RRRAA in its HVR3. Other alleles shown to be negatively associated with RA contain radical substitutions in HVR3 compared to the QRRAA template. The acidic residues which offer protection from RA lie in the wall of the antigen-binding cleft, and may prevent the binding of a pathogenic peptide. The implication is that either the QRRAA motif is required to bind and present the putative rheumatoid antigen to class II restricted T cells which then initiate an inflammatory response in the joint, or that the presence of the motif results in a different T cell receptor repertoire because of thymic selection.[12]

Although several alleles confer enhanced risk for development of RA, they are not equivalent, with important differences between ethnic groups and between subsets of patients. In northern Europe DRB1*0401 is the most strongly associated allele[8] whereas in Japanese, Jewish and Greek populations where DRB1*0401 is relatively uncommon, DRB1*0405 is the most common subtype among RA patients.[13-15] Conversely, where the alleles comprising the 'shared epitope' are rare, for example Nigeria, so is RA.[16]

Some recent studies have highlighted limitations of the shared epitope hypothesis. Several studies of mild, early RA in the community have suggested that the presence of the shared epitope is associated with disease severity rather than incidence.[17-19] A recent case control study has shown that the risk of RA in 'compound heterozygote' men who possess two different shared epitope alleles (*0401/*0404) is 90 times that in shared epitope negative individuals, but is more than double this in men under 30 with severe disease.[20] Such an interaction cannot be explained as a dosage effect since the DRB1*0401/0401 genotype is not associated with an equivalent risk.

INSULIN DEPENDENT DIABETES MELLITUS Twelve chromosomal regions have been implicated in the development of IDDM, and it is clear that it is a polygenic disease. The major disease locus (IDDM1) is found in the MHC, and is thought to be responsible for 35% of the observed familial clustering.[21] However, affected identical twin pairs have a risk of developing the disease of only 36%,[22] showing that environmental factors are important.

Early serological studies showed a positive association between IDDM and HLA B8, B15, DR3 and DR4, and a negative association or protective role for DR2. More recent molecular studies have shown stronger associations with other class II region antigens. The ability of DQB1 sequence variation to subdivide the DR2, DR4 and DRw6 haplotypes into susceptible and non-susceptible subtypes focused attention on the role of DQB1 in IDDM predisposition.[3] Todd[23] first outlined the hypothesis that DQB1 alleles with an aspartate at position 57 offer protection from IDDM, whereas alleles with a neutral position 57 residue lend an increased susceptibility. By predicting the structure of HLA-DQ on the basis of the crystal structure of HLA-DR1, this protection would appear to be rendered by the formation of a salt bridge across the end of the HLA antigen-binding cleft, which may prevent the binding and presentation of a pathogenic peptide.[5]

Although the IDDM association of a haplotype can be predicted to some extent by the charge of the polymorphic residue at position 57, not all DQB1 alleles with an aspartate at position 57 are equally protective. On the other hand, DR7, which is not associated with IDDM, has an alanine at position 57 of DQB1.[23] DQA alleles with an arginine at position 52 also appear to contribute susceptibility. Known associations of IDDM are with DQA1*0301 and *0302, DQB1*0201, *0302, DRB1*0301, *0401, and *0404, B8, B15, DR3 and 4. Studies in Caucasians have shown that Arg52+ DQ α/Asp57-DQβ heterodimers, in *cis* or *trans*, are strongly associated with susceptibility to IDDM.[24]

Again, one useful approach to understanding disease susceptibility loci has been to study the disease in a variety of ethnic groups where the haplotypes carrying susceptibility alleles may be distinct.

Numerous studies have been performed confirming the negative association with aspartate 57 in, for example, New Zealanders,[25] American[26] and North African[27] Blacks and Sardinians.[28] However, the association appears to break down in Japanese populations,[29] where there is a positive association with DQB1*0303, which contains Asp 57. DQB1*0303 is identical to *0302 except that the former has Asp 57. Since both are susceptibility loci in Japanese, it would appear that an Asp 57 positive allele confers susceptibility to IDDM when the whole molecule of the DQβ chain is similar to other susceptible DQβ chains. Clearly, in this regard, Todd's original hypothesis is found wanting.

COELIAC DISEASE Coeliac disease has an important genetic component, and first degree relatives of a patient have a 10% risk of being affected. There is a high concordance of around 70% in monozygotic twins although the number of twins studied is small. Early studies established strong associations with HLA-B8, DR3, DR7 and DQ2.[30] There is an HLA-DP1 association, but this is a consequence of the extended haplotype HLA-B8 DR3 DQ2.[31] Population studies have shown that the primary association is with both the DQA1*0501 and the DQB1*0201 alleles.[32]

One explanation for this population association is that a specific DQ α/β heterodimer encoded in a *cis* arrangement on DR3 haplotypes may be responsible for the observed DR associations, and is the primary disease association.[33] The α/β heterodimer is also encoded in a *trans* arrangement on DR5,7 genotypes and is more common in CD populations in Southern Europe.[34] Either of these heterodimer conformations is perhaps capable of presenting the pathogenic peptide to class II restricted T cells, whereas alleles not associated with CD are unable to present such a peptide.

Confirmation of the susceptibility locus has come from transracial mapping, showing that the association holds good in Italy[34] and Australia,[35] among others. Although Ashkenazi Jews have the expected excess of DQA1*0501, DQB1*0201, 30% of coeliac patients in this group do not carry these alleles, arguing against a single unifying hypothesis.[36] Approximately 10% of CD patients develop T cell lymphoma of the small intestine. A recent study showed that although the majority of patients with enteropathy-associated T cell lymphoma (EATCL) shared a similar HLA-DQ background as the uncomplicated patients, DR3,3 and DR3,7 genotypes were reduced and DR3,4 increased in EATCL.[37]

FELTY'S SYNDROME Felty's syndrome is an uncommon but severe variant of RA which develops in approximately 1% of RA patients, characterised by the association of arthritis with splenomegaly and neutropenia. There is a strong genetic component to Felty's syndrome, with 93% of the patients HLA DR4+ve, compared with 60-70% of RA patients and approximately 30% of controls.[38,39] In addition, there is an excess of DR4 homozygotes, particularly *0401/*0404 compound heterozygotes, giving a relative risk for disease development of 49. The risk associated with *0401 depends on the other allele present, with DRB1*0101 or *0404 homozygosity giving rise to a much lower relative risk.[39,40] Hence, whereas susceptibility to RA is strongly linked to the conserved HLA-DR beta epitope QRRAA associated with several DRB1 alleles, it is primarily the *0401 allele which is associated with progression to Felty's syndrome or severe RA.

The mechanism by which the two alleles give rise to this apparent multiplication of susceptibility is not clear. There may be alterations in thymic T cell receptor repertoire selection. Alternatively, it could be that the presence of two subtly different MHC molecules on the same cell surface results in the presentation of the same peptide to the incoming T cell, but with slightly differing orientations. This might either increase the T cell response by intracellular mechanisms, leading to more aggressive immune response with bystander damage, or reduce clearance of a pathogen from the tissues. RA is thus an example of a disease in which the examination of a severe, albeit rare, subtype can allow a closer examination of genetic susceptibility. This may facilitate the development of novel immunotherapies of relevance to the more common disease.

JUVENILE RHEUMATOID ARTHRITIS Likewise, patients with juvenile rheumatoid arthritis express the rare HLA DR 4 heterozygous phenotype not seen in unselected adult RA patients, but only in the more severe subtypes.[41] In addition, these patients have an increase in DPB2.1 allele not seen in adults,[42] independent of DR4 status. Hence it may be rewarding to examine the HLA associations of paediatric disease, since these apparently more severe subtypes can shed useful light on susceptibility loci in the more common adult diseases.

Limitations of HLA in disease susceptibility: role of other loci

Statistical models have been applied to the available data on HLA associated diseases, allowing a single locus model of disease susceptibility to be rejected for some. For example, the HLA-linked locus for CD accounts for only a five-fold increase in risk to siblings, whereas the observed risk is sixty-fold. Likewise, for IDDM the HLA-linked locus accounts for a three-fold increase in risk, whereas the observed value is fifteen-fold.[43] As discussed before, there are now twelve distinct chromosomal regions known to impart susceptibility to IDDM, although the major one lies in the MHC.

In RA, a number of groups have looked at regions and candidates within the MHC other than class I and II, and it appears unlikely that Transporter associated with Antigen Processing (TAP) polymorphisms confer susceptibility independent of DRB1.[44] However, the uncommon allele of Tumour Necrosis Factor α (TNF α) is found at three times the frequency of the normal Anglo Saxon population in patients with RA,[45] although this is not accompanied by an increase in production of TNF α protein. It is possible that this polymorphism modifies the disease course although the mechanism is not clear and it may represent a genetic marker for another locus. Indeed, the genome screen currently being undertaken at Oxford has shown preliminary evidence of an association between RA and a region close to, but not at, the TNF locus (Paul Wordsworth, personal communication). This finding awaits confirmation in a larger group of families.

For IDDM, statistical models have shown that the presence of non-Asp 57 in the DQβ chain is insufficient to explain the HLA component of IDDM predisposition.[46] A number of studies have looked elsewhere in the MHC, suggesting no independent role for TAP,[47] TNF[48] or heat shock protein[49] polymorphisms in IDDM predisposition. One recent study looked at LMP2 and LMP7, which are two subunits of proteasomes thought to play a role in the generation of endogenous peptides for presentation by class I molecules to cytotoxic T cells. In Caucasians, a genomic polymorphism of LMP7 was found to be strongly associated with IDDM independent of HLA DQ and DR genes.[50]

In CD, studies to find other susceptibility loci within the MHC have been fewer, but there is no TAP- specific disease association with CD in Italians or Ashkenazi Jews.[51]

Function of HLA susceptibility loci

All three diseases discussed have significant associations with HLA class II loci. The only known functions of class II molecules are to present self antigens in the thymus in foetal life to allow deletion of self-reactive clones and to shape the T cell receptor (TCR) repertoire, and to present self or foreign antigens in later life to CD4 'helper' T cells. The inference is that either subjects with these alleles fail to delete self-reactive T cells efficiently in development, or that they have an enhanced ability to present pathogenic self or foreign antigens to T cells, thus eliciting an inflammatory response. A third possibility is that these class II molecules are unable to present pathogenic peptides efficiently to T cells, resulting in a failure of elimination of the pathogen and subsequent disease.

For RA, support for the idea that the primary effect was on repertoire selection appeared to come from a study of patients and their mothers. This suggested that patients with RA who were negative for DR4 tended to have DR4+ mothers, the non-inherited allele therefore apparently affecting the foetal TCR repertoire.[52] However, a more recent study of the Arthritis and Rheumatism Council's National Repository of RA families found no such association.[53]

Evidence for the alternative, that susceptibility arises from an enhanced ability (or failure) to bind and present pathogenic peptides, has been sought in a number of approaches. One group recently purified HLA DR molecules from the spleen of a patient with Felty's syndrome.[54] The endogenous peptides were then isolated and sequenced. A variety of human proteins was found, notably albumin, and a synthesised peptide corresponding to residues 106-120 of albumin bound to HLA DR4 Dw4. Although the findings are interesting, it is difficult to see how this protein could give rise to the inflammatory changes seen in RA, which are largely restricted to tissues bearing collagen.

Two groups have employed site-directed mutagenesis on the DRB1*0401 gene. One found that single amino acid substitutions at position 71 resulted in changes in the portion of peptide selected that interacts with the 67-74 area.[55] The other found that similar changes at position 71 had a critical role in determining T cell responses to a given peptide, even though the binding affinity of the peptide was not significantly altered.[56]

Thus, there appears to be some persuasive evidence accumulating that changes in ability to present pathogenic peptide may explain the

HLA associations of these diseases. The development of mice transgenic for HLA DR or DQ molecules and human CD4 in the future should permit more extensive testing of the antigen presenting capabilities of these molecules.[57]

One interesting alternative hypothesis was recently outlined by Chella David's group.[58] Working from mouse models of arthritis, they proposed that HLA DQ was the arthritogenic peptide-presenting molecule, and that the HLA DRB1 locus was protective against RA. The proposed mechanism is that DRβ peptides are presented by HLA DQ molecules in the thymus, positively selecting certain TCR Vβ bearing cells. These T cells mature as CD4 cells with T helper 2 (Th2) type characteristics, and remain anergised in the periphery. When the immune system is activated by foreign antigens, specific Th1 cells and eventually DRβ-specific anergised Th2 cells are activated. Through antigenic mimicry, the Th1 cells become autoreactive. The Th2 cells then suppress the Th1 cells; therefore, depending on whether the DRB1 locus for that individual is permissive or protective, clinical disease will or will not ensue. This is a stimulating idea, and gives rise to a number of hypotheses which will doubtless be tested in the near future.

Conclusions

The work discussed above has improved our understanding of the nature of the genetic predisposition to autoimmune disease dependent upon the HLA region. Single loci responsible for the associations have been proposed, and tested at the population level. At the molecular level, our understanding of the interaction of the susceptibility-conferring HLA molecule, pathogenic peptide and T cell receptor has increased to the point where the goal of specific immunotherapies for these diseases appears more realistic. At the same time, the limitations of HLA in explaining these diseases has been recognised, and the role of genes within and without the MHC has been explored.

References

1 Trowsdale J, Campbell RD. Complexity in the major histocompatibility complex. Eur J Immunogenetics 1992;19(1/2):45-55.

2 Sloan VS, Cameron P, Porter G, Gammon M, Amaya M, Mellins E, et al.

Mediation by HLA-DM of dissociation of peptides from HLA-DR. Nature 1995;375:802-6.
3 Thomson G. HLA disease associations: models for the study of complex human genetic disorders. Crit Rev Clin Lab Sci 1995;32(2) 183-219.
4 Bjorkman PJ, Saper MA, Samraoui B, Bennett WS, Strominger JL, Wiley DC. The foreign antigen binding site and T cell recognition regions of class I histocompatibility antigens. Nature 1987;329:512-8.
5 Brown JH, Jardetzky TS, Gorga JC, Stern LJ, Urban RG, Strominger JL, et al. Three-dimensional structure of the human class II histocompatibility antigen HLA-DR1. Nature 1993;364:33-9.
6 Hughes AL, Nei M. Pattern of nucleotide substitution at major histocompatibility complex class I loci reveals overdominant selection. Nature 1988;335:167-70.
7 Hughes AL, Nei M. Nucleotide substitution at major histocompatibility complex class II loci: evidence for overdominant selection. Proc Natl Acad Sci USA 1989;86:958-62.
8 Wordsworth BP, Lanchbury JSS, Sakkas LI, Welsh KI, Panayi GS, Bell JI. HLA-DR4 subtype frequencies in rheumatoid arthritis indicate that DRB1 is the major susceptibility locus within the HLA class II region. Proc Natl Acad Sci USA 1989;86:10049-53.
9 Gao X, Olsen NJ, Pincus T, Stastny P. HLA-DR alleles with naturally occurring amino acid substitutions and risk for development of rheumatoid arthritis. Arthr Rheum 1990;33:939-46.
10 Tan PL, Farmiloe S, Roberts M, Geursen A, Skinner MA. HLA-DR4 subtypes in New Zealand Polynesians. Predominance of Dw13 in the healthy population and association of Dw15 with rheumatoid arthritis. Arthr Rheum 1993;36(1):15-9.
11 Gregerson P, Silver J, Winchester R. The shared epitope hypothesis. An approach to understanding the molecular genetics of susceptibility to rheumatoid arthritis. Arthr Rheum 1987;30:1205-13.
12 Gregersen PK. HLA associations with rheumatoid arthritis: a piece of the puzzle. J Rheumatol 1992;32(suppl.):7-11.
13 Watanabe Y, Tokunga K, Matsuki K, Takeuchi F, Matsuta K, Maeda H, et al. Putative amino acid sequence of HLA-DRB chain contributing to rheumatoid arthritis. J Exp Med 1989;169:2263-8.
14 Gao X, Serjeantson SW. Diversity in HLA-DR4-related DR,DQ haplotypes in Australia, Oceania, and China. Hum Immunol 1991;32(4):269-76.
15 Boki KA, Panayi GS, Vaughan RW, Drosos AA, Moutsopoulos HM, Lanchbury JS. HLA class II sequence polymorphisms and susceptibility to rheumatoid arthritis in Greeks. The HLA-DR beta shared-epitope hypothesis accounts for the disease in only a minority of Greek patients. Arthr Rheum 1992;35(7):749-55.
16 Silman AJ, Ollier W, Holligan S, Birrell F, Adebajo A, Asuzu MC, et al. Absence of rheumatoid arthritis in a rural Nigerian population. J Rheumatol 1993;20(4):618-22.
17 Deighton CM, Kelly PJ, Walker DJ. Linkage of rheumatoid arthritis with HLA. Ann Rheum Dis 1993;52(9):638-42.

18 Thomson W, Pepper L, Payton A, Carthy D, Scott D, Ollier W, et al. Absence of an association between HLA-DRB1*04 and rheumatoid arthritis in newly diagnosed cases from the community. Ann Rheum Dis 1993;52(7):539-41.

19 Weyand CM, Hicok KC, Conn DL, Goronzy JJ. The influence of HLA-DRB1 genes on disease severity in rheumatoid arthritis. Ann Intern Med 1992;117(10):801-6.

20 MacGregor A, Ollier W, Thomson W, Jawaheer D, Silman A. HLA-DRB1*0401/0404 genotype and rheumatoid arthritis: increased association in men, young age at onset, and disease severity. J Rheumatol 1995;22(6):1032-6.

21 Cordell HJ, Todd JA. Multifactorial inheritance in type 1 diabetes. Trends Genetics 1995;11(12):499-504.

22 Olmos P, A'Hern R, Heaton DA, Millward BA, Risley D, Pyke DA, et al. The significance of the concordance rate for type 1 (insulin-dependent) diabetes in identical twins. Diabetologia 1988;31(10):747-50.

23 Todd JA, Bell JI, McDevitt HO. HLA-DQ beta gene contributes to susceptibility and resistance to insulin-dependent diabetes mellitus. Nature 1987;329(6140):599-604.

24 Khalil I, Deschamps I, Lepage V, al-Daccak R, Degos L, Hors J. Dose effect of *cis*- and *trans*-encoded HLA-DQ alpha beta heterodimers in IDDM susceptibility. Diabetes 1992;41(3):378-84.

25 Forbes LV, Brown LJ, Scott RS. HLA-DQ beta typing and non-Asp57 alleles in IDDM and nondiabetic subjects in New Zealand. Diabetes Care 1993;16(8):1179-83.

26 Fernandez-Vina M, Ramirez LC, Raskin P, Stastny P. Genes for insulin-dependent diabetes mellitus (IDDM) in the major histocompatibility complex (MHC) of African-Americans. Tissue Antigens 1993;41(2):57-64.

27 Djoulah S, Khalil I, Beressi JP, Benhamamouch S, Bessaoud K, Deschamps I, et al. The HLA-DRB1*0405 haplotype is most strongly associated with IDDM in Algerians. Eur J Immunogenetics 1992;19(6):381-9.

28 Cucca F, Muntoni F, Lampis R, Frau F, Argiolas L, Silvetti M, et al. Combinations of specific DRB1, DQA1, DQB1 haplotypes are associated with insulin-dependent diabetes mellitus in Sardinia. Hum Immunol 1993;37(2):85-94.

29 Ikegami H, Kawaguchi Y, Yamato E, Kuwata S, Tokunaga K, Noma Y, et al. Analysis by the polymerase chain reaction of histocompatibility leucocyte antigen-DR9-linked susceptibility to insulin-dependent diabetes mellitus. J Clin Endocrinol Metab 1992;75(5):1381-5.

30 Ellis A, Taylor CJ, Dillon-Remmy M, Woodrow JC, McConnell RB. HLA-DR typing in coeliac disease: evidence for genetic heterogeneity. Br Med J (Clin Res Ed) 1984;289(6458):1571-3.

31 Hall MA, Lanchbury JS, Bolsover WJ, Welsh KI, Ciclitira PJ. Celiac disease is associated with an extended HLA-DR3 haplotype which includes HLA-DPw1. Hum Immunol 1990;27(3):220-8.

32 Sollid LM, Markussen G, Ek J, Gjerde H, Vartdal F, Thorsby E. Evidence for a primary association of coeliac disease to a particular HLA-DQ α/β heterodimer. J Exp Med 1989;169:345.

33 Hall MA, Lanchbury JS, Lee JS, Welsh KI, Ciclitira PJ. HLA-DQ2 second-

domain polymorphisms may explain increased *trans* -associated risk in celiac disease and dermatitis herpetiformis. Hum Immunol 1993;38(4):284-92.

34 Tighe MR, Hall MA, Barbado M, Cardi E, Welsh KI, Ciclitira PJ. HLA class II alleles associated with celiac disease susceptibility in a southern European population. Tissue Antigens 1992;40(2):90-7.

35 Hetzel PA, Bennett GD, Sheldon AB, Propert DN, Brown R, Hay JA, et al. Genetic markers in Australian Caucasian subjects with coeliac disease. Tissue Antigens 1987;30(1):18-22.

36 Tighe MR, Hall MA, Ashkenazi A, Siegler E, Lanchbury JS, Ciclitira PJ. Celiac disease among Ashkenazi Jews from Israel. A study of the HLA class II alleles and their associations with disease susceptibility. Hum Immunol 1993;38(4):270-6.

37 Howell WM, Leung ST, Jones DB, Nakshabendi I, Hall MA, Lanchbury JS, et al. HLA-DRB, -DQA, and -DQB polymorphism in celiac disease and enteropathy-associated T-cell lymphoma. Common features and additional risk factors for malignancy. Hum Immunol 1995;43(1):29-37.

38 Campion G, Maddison PJ, Goulding N, James I, Ahern MJ, Watt I, et al. The Felty syndrome: a case-matched study of clinical manifestations and outcome, serologic features, and immunogenetic associations. Medicine (Baltimore) 1990;69(2):69-80.

39 Lanchbury JS, Jaeger EE, Sansom DM, Hall MA, Wordsworth P, Stedeford J, et al. Strong primary selection for the Dw4 subtype of DR4 accounts for the HLA-DQw7 association with Felty's syndrome. Hum Immunol 1991;32(1):56-64.

40 Wordsworth P, Pile KD, Buckely JD, Lanchbury JS, Ollier B, Lathrop M, et al. HLA heterozygosity contributes to susceptibility to rheumatoid arthritis. Am J Hum Genet 1992;51(3):585-91.

41 Nepom BS, Nepom GT, Mickelson E, Schaller JG, Antonelli P, Hansen JA. Specific HLA-DR4-associated histocompatibility molecules characterize patients with seropositive juvenile rheumatoid arthritis. J Clin Invest 1984;74:287-91.

42 Begovich AB, Bugawan TL, Nepom BS, Klitz W, Nepom GT, Erlich HA. A specific HLA-DP beta allele is associated with pauciarticular juvenile rheumatoid arthritis but not adult rheumatoid arthritis. Proc Natl Acad Sci USA 1989;86(23):9489-93.

43 Risch N. Assessing the role of HLA-linked and unlinked determinants of disease. Am J Hum Genet 1987;40(1):1-14.

44 Marsal S, Hall MA, Panayi GS, Lanchbury JS. Association of TAP2 polymorphism with rheumatoid arthritis is secondary to allelic association with HLA-DRB1. Arthr Rheum 1994;37(4): 504-13.

45 Danis VA, Millington M, Hyland V, Lawford R, Huang Q, Grennan D. Increased frequency of the uncommon allele of a tumour necrosis factor alpha gene polymorphism in rheumatoid arthritis and systemic lupus erythematosus. Dis Markers 1995;12(2):127-33.

46 Robinson WP, Barbosa J, Rich SS, Thomson G. Homozygous parent affected sib pair method for detecting disease predisposing variants: application to insulin dependent diabetes mellitus. Genet Epidemiol 1993;10(5):273-88.

47 Nakanishi K, Kobayashi T, Murase T, Kosaka K. Lack of association of the transporter associated with antigen processing with Japanese insulin-dependent diabetes mellitus. Metab Clin Exp 1994;43(8):1013-7.

48 Cox A, Gonzalez AM, Wilson AG, Wilson RM, Ward JD, Artlett CM, et al. Comparative analysis of the genetic associations of HLA-DR3 and tumour necrosis factor alpha with human IDDM. Diabetologia 1994;37(5):500-3.

49 Pugliese A, Awdeh ZL, Galluzzo A, Yunis EJ, Alper CA, Eisenbarth GS. No independent association between HSP70 gene polymorphism and IDDM. Diabetes 1992;41(7):788-91.

50 Deng GY, Muir A, Maclaren NK, She JX. Association of LMP2 and LMP7 genes within the major histocompatibility complex with insulin-dependent diabetes mellitus: population and family studies. Am J Hum Genet 1995;56(2):528-34.

51 Tighe MR, Hall MA, Cardi E, Ashkenazi A, Lanchbury JS, Ciclitira PJ. Associations between alleles of the major histocompatibility complex-encoded ABC transporter gene TAP2, HLA class II alleles, and celiac disease susceptibility. Hum Immunol 1994;39(1):9-16.

52 Ten Wolde S, Breedveld FC, De Vries RR, D'Amaro J, Rubenstein P, Schreuder GM, et al. Influence of non-inherited maternal HLA antigens on occurrence of rheumatoid arthritis. Lancet 1993;341(8839):200-2.

53 Silman AJ, Hay EM, Worthington J, Thomson W, Pepper L, Davidson J, et al. Lack of influence of non-inherited maternal HLA-DR alleles on susceptibility to rheumatoid arthritis. Ann Rheum Dis 1995;54(4):311-3.

54 Gordon RD, Young JA, Rayner S, Luke RW, Crowther ML, Wordsworth P, et al. Purification and characterization of endogenous peptides extracted from HLA-DR isolated from the spleen of a patient with rheumatoid arthritis. Eur J Immunol 1995;25(5): 1473-6.

55 Hammer J, Gallazzi F, Bono E, Karr RW, Guenot J, Valsasnini P, et al. Peptide binding specificity of HLA-DR4 molecules: correlation with rheumatoid arthritis association. J Exp Med 1995;181(5):1847-55.

56 Signorelli KL, Watts LM, Lambert LE. The importance of DR4Dw4 beta chain residues 70, 71, and 86 in peptide binding and T cell recognition. Cell Immunol 1995;162(2):217-24.

57 Fugger L, Michie SA, Rulifson I, Lock CB, McDevitt GS. Expression of HLA-DR4 and human CD4 transgenes in mice determines the variable region beta-chain T-cell repertoire and mediates an HLA-DR-restricted immune response. Proc Natl Acad Sci USA 1994;91(13):6151-5.

58 Zanelli E, Gonzalez-Gay MA, David CS. Could HLA-DRB1 be the protective locus in rheumatoid arthritis? Immunol Today 1995;16(6):274-8.

J. WAYNE STREILEIN

Ocular immune privilege and autoimmunity

Introduction

The primary function of the eye is to receive accurate light images and to transduce and transmit them faithfully to the brain. Carrying out this primary function requires the existence of the microanatomical structure known as the visual axis. Maintenance of this precise structure is essential for the preservation of vision. The visual axis is therefore vulnerable to anatomic distortions, especially those caused by intense inflammation. In that regard, certain immune effector modalities, particularly delayed hypersensitivity and complement fixing antibodies, mediate protection against microorganisms primarily by enlisting non-specific inflammatory mediators and processes. As a consequence, full intraocular expression of delayed hypersensitivity and complement fixation (which may very well lead to the elimination of an intraocular pathogen) may unavoidably cause disruption of the visual axis and lead to the loss of vision. For the past 20 years our laboratory has been interested in trying to understand how the eye deals with this dilemma. Our studies on ocular immune privilege have begun to elucidate some of the mechanisms involved.

Ocular immune privilege

It has been known for more than 100 years that the anterior chamber is an immunologically privileged site.[1] Since that description, the literature has been, if anything, confused concerning what immune privilege actually means. Based on work from our laboratory and others, ocular immune privilege can be defined operationally as follows:[2,3] foreign tissues placed in the anterior chamber, the vitreous cavity and the subretinal space experience extended (even indefinite) survival, whereas similar tissues placed subcutaneously (at conventional sites) are promptly rejected. Unlike the original presumptions regarding the basis for immune privilege, intraocular graft survival is actually maintained for long periods by active rather than passive processes. Moreover, privilege extended to foreign tissues placed in the eye may not necessarily be absolute. That is to say, some grafts may experience extended survival compared to conventional sites, but they may not be tolerated within the eye indefinitely. The fact that privilege is not 'absolute' underscores that the phenomenon is actively maintained and regulated by the immune system. This realization holds the promise that survival of intraocular grafts and tissues should be able to be manipulated in both positive and negative directions.

There is now strong evidence to indicate that immune privilege results from alterations in both the induction and the expression of immunity to foreign tissues placed within the eye. Finally, it seems reasonable to speculate that immune privilege in the eye is an evolutionary adaptation designed to limit the intraocular expression of immunogenic inflammation as a means of preserving vision.

Features of the eye pertinent to immune privilege

Several features of the eye appear to be critical to the existence of ocular immune privilege.[4] First, there is an anatomical and physiological barrier that isolates the eye from the contents of the blood. This blood-ocular barrier regulates the ability of molecules and cells in the blood to gain access to the eye and its internal compartments. Second, the eye has a relative-to-absolute lack of lymphatic drainage pathways. Realization of this fact more than 40 years ago led immunologists at the time to think that immune privilege resulted from sequestration of antigens in a site that the immune system could not detect. However, aqueous humor flows through the trabecular meshwork directly into the venous drainage of the eye, and therefore intraocular contents, including antigenic material, can gain access to the systemic immune apparatus through an intravenous route of delivery. Third, recent experiments have formally identified large numbers of bone marrow-derived, even MHC class II-expressing cells in the iris, ciliary body, choroid and retina. Direct evaluation of the antigen-presenting capabilities of these cells has revealed them to be deficient – in a conventional sense.[5,6] Thus, the eye contains putative professional antigen-presenting cells that display atypical functional properties. Fourth, ocular fluids, especially the aqueous humor, have been demonstrated to contain a rather wide range of factors (cytokines, growth factors, neuropeptides) that individually, and in the aggregate, suppress immunogenic inflammation and antibody-dependent complement activation.[7]

It is worth emphasizing that recent studies have demonstrated that immune privilege, at least in the anterior chamber of the eye, results, on the one hand, from a modification of the afferent limb of immune induction by which the immune system becomes aware of and recognizes antigens placed within this ocular site.[2-4] On the other hand, the ability of conventional immune effectors to express themselves within the eye is markedly modified by the intraocular immunosuppressive microenvironment.[8] Thus, immune privilege results from stereotypic modifications in both the induction and the expression of immunity to ocular antigens.

Anterior chamber-associated immune deviation (ACAID) – Role in ocular immune privilege

Over the past 20 years, several laboratories have dedicated their efforts to understanding the modifications in induction of immunity that take place when antigens are injected into the anterior chamber of the eye.[2-4] The unusual systemic immunity that is observed in response to ocular antigens has been termed Anterior Chamber Associated Immune Deviation (ACAID). This phenomenon is defined operationally as follows: antigens placed in the anterior chamber evoke a selective, systemic immune deficiency in which the T cells that mediate delayed hypersensitivity and the B cells that secrete complement fixing antibodies are missing, whereas other antigen-specific immune effectors, such as CD8+ T cells and B cells that secrete non-complement fixing antibodies, are intact and stimulated. Based on numerous studies, ACAID has been found to display certain features which are distinctive and which set it apart from other types of systemic immune responses to antigen. First, ACAID can be induced by a wide variety of physicochemically distinct antigens, ranging from histocompatibility antigens to tumor-specific antigens, viral encoded antigens, haptens, and heterologous soluble protein antigens. Except for antigens of high immunogenicity (MHC class

I molecules, tumor antigens induced by UV irradiation), ACAID is a very long-lasting phenomenon – apparently lasting for the life of the animal. When an antigen is injected into a subcutaneous compartment and into the anterior chamber simultaneously, the systemic immune response resembles that of ACAID rather than conventional immunity. This indicates not only that ACAID is dominant in this setting, but it implicates potent regulatory forces as being operative. Perhaps the most dramatic expression of this fact is that ACAID can even be imposed on animals previously sensitized to the antigen in question.[9] Although ACAID was originally described as a deviant immune response to antigens placed in the anterior chamber of the eye, a similar phenomenon exists when antigen is placed in other intraocular compartments: the vitreous cavity and subretinal space.[10] Finally, it should be mentioned that ACAID has not only been demonstrated in mice and rats, but is also a characteristic of primate eyes.[11]

The role of anterior chamber-associated immune deviation in the pathogenesis of uveitis

Although several experimental models of uveitis have been described, perhaps the most experience has been gained with the uveoretinitis induced by systemic immunization with soluble protein antigens derived from the mammalian retina – experimental autoimmune uveitis. Our laboratory has been interested in the relationship of ACAID (and by implication immune privilege) to the pathogenesis of this experimental uveitis. We have explored three experimental questions: 1. Can ACAID be induced by the intraocular injection of retinal autoantigens? 2. Does ACAID to such antigens modify the expression of experimental autoimmune uveoretinitis, and if so, by what mechanism? 3. Can antigen-specific ACAID be used to prevent or suppress autoimmune uveoretinitis? A summary of experimental results bearing on these questions follows.

Using interphotoreceptor retinal binding protein (IRBP) as the prototypic retinal autoantigen, we have demonstrated that intraocular injection of this antigen into BALB/c mice induces IRBP-specific ACAID.[12] Mice that have received an initial injection of antigen through this route do not develop delayed hypersensitivity when immunized subsequently with IRBP in complete Freund's adjuvant. This indicates that endogenous antigens of the retina are capable of inducing the ACAID phenomenon.

The B10.A strain of mice not only develops delayed hypersensitivity when immunized with IRBP, but the vast majority of such mice develop uveoretinitis. Accordingly, we have pretreated B10.A mice with an intraocular injection of IRBP and then attempted to induce uveitis in these mice by administering to them a uveitogenic regimen of IRBP plus adjuvant. We discovered that B10.A mice developed ACAID when IRBP was injected into the anterior chamber and that the incidence of clinically detected uveitis in these mice was extremely low compared to positive controls. Moreover, among the few animals that displayed uveitis, the intensity was of a lower magnitude than that observed in positive control eyes. Thus, it would appear that a maneuver designed to prevent delayed hypersensitivity to an ocular autoantigen by injecting the antigen first into the anterior chamber, induces ACAID and prevents the induction and expression of EAU.

Spleens of mice with ACAID contain antigen-specific regulatory cells that interfere with both the induction and the expression of delayed hypersensitivity to the antigen in question.[13] We have inquired whether the suppressor cells found within spleens of animals pretreated with IRBP can be used in an adoptive transfer assay to prevent or suppress the expression of EAU in recipient mice. In these experiments, the adoptive transfer was carried out a) prior to immunization with the IRBP in uveitogenic form, b) after immunization but prior to the expression of intraocular inflammation, and c) after inflammation had developed in the eyes of recipient mice. Our findings indicated that spleen cells from mice with IRBP-specific ACAID can prevent the induction of EAU in mice with IRBP plus adjuvant.[12] More importantly, these regulatory cells can even quench the expression of inflammation in eyes of animals with extant disease. Studies of this type have enabled us to conclude that the induction of ACAID by retina-specific antigens prevents the induction of EAU, in part through the generation of suppressor cells that can inhibit intraocular immunogenic inflammation – even after it has been initiated.

ACAID as a potential therapy for established experimental autoimmune uveitis

As mentioned previously, ACAID can be induced by intraocular injection of antigen into animals already sensitized to the antigen in question.[9] This raises the possibility that ACAID to IRBP might be induced even after animals had received a uveitogenic regimen designed to elicit EAU. We have designed experiments to test two experimental questions. In the first instance, we inquired whether an anterior chamber injection of IRBP into IRBP-immune mice can induce ACAID. We discovered that IRBP resembles other soluble protein antigens in that animals originally sensitized to IRBP lose their delayed hypersensitivity reactivity to this antigen if they receive an intraocular injection of IRBP.[9]

It has recently been reported that ACAID is induced by mobile antigen-presenting cells that pick up antigen in the anterior chamber and migrate via the blood to the spleen.[14,15] Although these cells arise within the eye under physiological circumstances, we have been able to create the equivalent of an anterior chamber *in vitro* by incubating peritoneal exudate cells in culture medium containing transforming growth factor beta.[16] When such cells are exposed to TGFβ and pulsed with soluble protein antigen, they acquire ACAID-inducing properties which are revealed when the cells are subsequently washed and injected intravenously into naive recipients. These recipients are impaired in their ability to develop delayed hypersensitivity when immunized subsequently to the antigen in question. We therefore asked whether we could create an IRBP-specific ACAID-inducing signal *in vitro* and then use this cell-bound signal *in vivo* to limit the expression of EAU. Our results indicated that an ACAID-inducing signal with specificity for IRBP can be created by incubating peritoneal exudate cells with IRBP and TGFβ *in vitro*.[9] Moreover, when cells prepared in this way were injected intravenously into B10.A mice with extant uveitis secondary to immunization with IRBP, the intraocular inflammation subsided over the next several days. Since no similar reduction in inflammation occurred in control eyes, we conclude that an ACAID-inducing signal generated completely *in vitro* may offer a therapeutic approach to reducing autoimmune intraocular inflammation.

Summary and implications

Our experiments indicate that the strategies used by the eye to create an immune privileged site can be used to advantage with regard to the induction, modification and prevention of experimental autoimmune uveitis. Using the strategy of ACAID, ocular antigens have been used to prevent ocular autoimmune disease and to suppress ongoing and existing disease. This strategy might be considered for its possible clinical applications in the future. It should be apparent that the power of immune privilege to down-regulate experimental autoimmune disease may relate to a primary function for immune privilege in physiology.

References

1 Medawar P. Immunity to homologous grafted skin. III. The fate of skin homografts transplanted to the brain, to subcutaneous tissue, and to the anterior chamber of the eye. Br J Exp Pathol 1948; 29:58-69.

2 Niederkorn JY. Immune privilege and immune regulation in the eye. Adv Immunol 1990; 48:191-226.

3 Streilein JW. Immune regulation and the eye: A dangerous compromise. FASEB J 1987; 1:199-208.

4 Streilein JW. Immune privilege as the result of local tissue barriers and immunosuppressive microenvironments. Curr Opin Immunol 1993; 5:428-32.

5 Williamson JSP, Bradley D, Streilein JW. Immunoregulatory properties of bone marrow derived cells in the iris and ciliary body. Immunology 1989; 67:96-102.

6 Streilein JW, Bradley D. Analysis of immunosuppressive properties of iris and ciliary body cells and their secretory products. Invest Ophthalmol Vis Sci 1991; 32:2700-10.

7 Streilein JW, Wilbanks GA, Cousins SW. Immunoregulatory mechanisms of the eye. J Neuroimmunol 1992; 39:185-200.

8 Cousins S, Trattler W, Streilein JW. Immune privilege and suppression of immunogenic inflammation in the anterior chamber of the eye. Curr Eye Res 1991; 10:287-97.

9 Kosiewicz MM, Okamoto S, Miki S, Ksander BR, Shimizu T, Streilein JW. Imposing deviant immunity on the presensitized state. J Immunol 1994; 153:2962-73.

10 Jiang LQ, Jorquera M, Streilein JW. Subretinal space and vitreous cavity as immunologically privileged sites for retinal allografts. Invest Ophthalmol Vis Sci 1993; 34:3347-54.

11 Eichorn M, Horneber M, Streilein JW, Lutjen-Drecoll E. Anterior chamber associated immune deviation elicited via primate eyes. Invest Ophthalmol Vis Sci 1993; 34:2926-30.

12 Hara Y, Caspi RR, Wiggert B, Chan CC, Wilbanks GA, Streilein JW. Suppression of experimental autoimmune uveitis in mice by induction of Anterior Chamber Associated Immune Deviation with interphotoreceptor retinoid binding protein. J Immunol 1992; 148:1685-92.

13 Wilbanks GA, Streilein JW. Characterization of suppressor cells in Anterior Chamber Associated Immune Deviation (ACAID) induced by soluble antigen. Evidence of two functionally and phenotypically distinct T-suppressor cell

populations. Immunology 1990; 71:383-9.
14 Wilbanks GA, Streilein JW. Studies on the induction of Anterior Chamber Associated Immune Deviation (ACAID). I. Evidence that an antigen-specific, ACAID-inducing, cell-associated signal exists in the peripheral blood. J Immunol 1991; 146:2610-7.
15 Wilbanks GA, Mammolenti MM, Streilein JW. Studies on the induction of Anterior Chamber Associated Immune Deviation (ACAID). II. Eye-derived cells participate in generating blood borne signals that induce ACAID. J Immunol 1991; 146:3018-24.
16 Wilbanks GA, Mammolenti MM, Streilein JW. Studies on the induction of Anterior Chamber Associated Immune Deviation (ACAID). III. Induction of ACAID depends upon intraocular transforming growth factor-ß. Eur J Immunol 1992; 22:165-73.

ORTWIN ROTT AND EVELYNE CASH

Viruses and the induction of autoimmune diseases

Introduction

There have been numerous suggestions that viral infections may be the trigger for the induction and/or perpetuation of autoimmune diseases, i.e. immunopathologic entities of currently unknown etiology, associated with immune responses against self-antigens. The potential of viruses to establish long-term (and sometimes life-long) persistent or latent infections renders these infectious agents a prime candidate to alter the proper functioning of the immune system and to induce disease in susceptible hosts. Despite intensive immune research, however, no unequivocal association between a given autoimmune disease and infection by a single type of virus has ever been established. This observation suggests that the clue for our understanding of how viruses induce autoimmunity might not be the analysis of immune cell interactions with a multitude of antigens encoded by a single type of virus, but rather the elucidation of host cell interactions with a particular group of viral proteins encoded by a multitude of viruses. In other words, the characteristics of a given viral protein as a *functional entity* for non-cognate interactions with the immune system might play a much more prominent role in autoimmunity/immunopathology than the unique character of the same viral protein as an *antigenic entity* for determining cognate interactions with immune cells (as illustrated in, for example, the classical concept of molecular mimicry).

In fact, the concept of particular groups of "immunopathologic" viral antigens which have very similar or nearly identical immunopathogenic properties, in spite of great differences in the primary structure of the viral proteins, is given strong support by a number of experimental approaches in recent years. The common basis of all of these at first glance rather heterogeneous experimental set-ups is the emphasis on functional similarities in immune interactions of isolated viral proteins. Examples of these groups of "immunopathologic" viral antigens include:

1. Stimulation of T lymphocytes by polyclonal-activating viral proteins, i.e. T cell 'superantigens'.
2. Stimulation of cytokines by regulatory viral proteins, i.e. retroviral encoded 'transactivating' antigens.
3. Stimulation of B lymphocytes by polyclonal-activating viral proteins, i.e. (T cell-dependent or -independent) B cell 'mitogenic' antigens.

This review will attempt to illustrate our current knowledge as to the possible implications of these diverse types of "immunopathologic" viral antigens on experimental and clinical autoimmunity.

T cell 'superantigens' and autoimmunity

Superantigens are a heterogeneous class of microbial proteins which bind specifically to T cell receptor (TCR) and MHC class II molecules.[1] Contrary to conventional protein antigens, where about $1/10^5$ T cells respond to a given peptide, viral and bacterial superantigens interact with up to 20% of the peripheral T cell pool after primary antigen encounter. The polyclonal T cell activation potential of superantigens is the reflection of a particular mode of molecular interaction of superantigens with the T cell antigen receptor. Superantigens (SAgs) do not act as peptides and do not seem to engage all the variable components of the TCR, but bind to a conserved region of the Vβ chain of the TCR.[2-5] This Vβ-specific activation mode of SAgs has led to numerous speculations and experimental approaches on the potential involvement of SAgs in autoimmune diseases with a suspected limitation in TCR Vß-heterogeneity among effector lymphocytes. In the following, some of these experimental approaches and the current knowledge as to the relevance of SAgs in human diseases will be outlined. Furthermore, as most of the experimental data were obtained in studies with bacterial SAgs (notably *Staphylococcus aureus* enterotoxins, SE), this group of SAg will be included in our discussion. For a more detailed information about the molecular basis for T cell responses to SAgs, a number of excellent, recently written reviews can be recommended.[2–5]

Before considering the potential impact of SAgs on the generation of pathologic (autoimmune) responses, the general features of T cell reactivity towards SAgs *in vitro* and *in vivo* should be called to mind. Like responses to conventional Ags, T cell responses to SAgs differ in the thymic- and post-thymic environment: whereas a T cell encounter with SAgs in the thymus leads to deletion, thereby ensuring tolerance towards superantigenic "self-antigens", confrontation of mature T cells with SAgs induces a proliferative activation.[6–9] In contrast to conventional Ags, however, this activation of peripheral T cells by SAg is *aberrant*, i.e. results in a subsequent state of Vß-specific unresponsiveness. This induction of Vß-specific unresponsiveness, first demonstrated by Rammensee et al.[6] for mouse mammary tumor virus (MMTV) encoded Mls SAg, is due to both deletion (physical elimination) and anergy (functional elimination) of the responding T lymphocytes. Although this is still very poorly understood, some groups have recently suggested that this anergy-inducing potential of SAgs might be the reflection of an alternative signal transduction pathway utilized by these agents.[10–14]

Anergy induction by SAgs implies that superantigenic proteins can functionally paralyze (pathogenic) TCR $\alpha\beta^+$ T cells *in vivo.* Thus, rather than positively participating in the specific induction of immune responses against nominal antigens, SAgs have been shown to "turn-off" proliferative responses to various peptide antigens in the respective Vß T cell subpopulation. The first study[15] of such a functional "paralysis" of an autoimmune response by SAgs was carried out in the widely used animal model of experimental autoimmune encephalomyelitis (EAE) in Lewis rats. In this particular rat strain, the pathogenic mechanism underlying EAE is the activation of myelin basic protein (MBP)-specific T_h1 effector cells, which belong almost exclusively to the Vß8.2 $CD4^+$ subset. Induction of EAE in Lewis rats could be specifically prevented by the administration of Vß8.2 activating, bacterial SAgs (SEE) prior to immunization with MBP.[15] Such an abrogation (or at least strong reduction) of autoimmune disease by pretreatment with the appropriate Vß-targeting bacterial SAg *in vivo* was subsequently extended to the mouse model of EAE.[16,17] Thus, SAgs can suppress autoimmune reactivities by anergizing potentially autoaggressive effector T cells expressing the appropriate Vß element(s).[15]

Do these findings refute any positive function of SAgs in the immune pathomechanism of T cell-mediated autoimmune diseases? Certainly not. Again from data obtained with bacterial SAgs, there is now clear evidence that one of the immediate effects of the administration of SAgs *in vivo* is the hyperexpression of a number of cytokines during the initial expansion phase of T cell reactivity. As shown by Bette et al.,[18] SAg-stimulated T cells produce a number of proinflammatory cytokines in a "burst-like" fashion, before being fated towards anergy. This "cytokine-burst" by SAg-activated T lymphocytes[18,19] provides the molecular basis for a well-documented form of immunopathology in man, the so-called toxic shock syndrome (TSS), observed in the course of severe infections with SAg-producing Gram-negative bacteria. Interestingly, the predominant cytokines released by (viral or bacterial) SAg-activated T cells belong to the T_h1-like group of lymphokines (IL-2, IFN-γ, TNF-α).[18,20] It is plausible to assume that this massive cytokine release by SAgs-activated T cells could lead to a (non-Vß-specific) reactivation of an already ongoing inflammatory process, including autoimmune responses. Indeed, SAg administration to already diseased animals has been shown to reactivate local tissue inflammation and clinical signs of autoimmunity: Using the murine model of EAE, the administration of various SEs with different Vß-specificities could "booster" the already ongoing process of central nervous system (CNS) inflammation[21,22] and result in an aggravation of clinical signs of CNS disease. Although not directly addressed, it is likely that T_h1-cytokine release played a major role in reactivation of EAE in these studies. An alternative mechanism was demonstrated in a recent report, where the SE-mediated induction of MBP-directed primary Vß8.2 T cell responses in naive Lewis rats was evaluated.[23] Interestingly, co-immunization of rats with MBP and a directly Vß8.2 activating SAg (SEE) could not induce autoreactive T cell responses, whereas co-immunization with a non-Vß8.2-activating SAg (SEA) resulted in MBP-specific autoreactivities of Vß8.2$^+$ T cells. Thus, SAgs seem to be able to induce primary T cell responses to autoantigens by a non-Vß-specific mechanism of "bystander" activation.[23]

In essence, the currently available experimental data have therefore addressed two different situations of T cell encounter with SAgs: In the case of murine retroviral-encoded, endogenous SAgs, SAg-reactive T lymphocytes are intrathymically deleted, thereby establishing a concomitant state of tolerance to a (superantigenic) self-antigen. Whether such a SAg-triggered shaping of the TCR repertoire could have any important consequences on the later induction of autoimmunity in the adult animal is questionable. In the case of T cell encounter with soluble, bacterial SAgs in the adult host, the SAg might under certain circumstances induce a primary T cell autoreactive immune response by an "innocent" bystander activation mechanism,[23] or reactivate an already ongoing inflammatory process by a SAg-mediated "cytokine-burst" in responding T cells.[18,19] In these two examples, the contribution to autoimmunity is only *indirectly* mediated by the SAg-activated Vß-subpopulation.

What remains is a third and certainly the most interesting scenario: The infection of the adult, non-tolerant host with a SAg-encoding virus, leading to a tissue-restricted, chronic expression of a SAg in the virus-infected organ(s). In such a scenario, local SAg-expression would probably not result in a massive and immediate deletion of circulating SAg-reactive T cells, but rather in the entertainment of a low-grade, chronic stimulation of SAg-reactive T cells at a restricted site in the organism. It might be expected that such a local, SAg-triggered inflammatory process could present a number of typical features of organ-specific "autoimmune" diseases. However, for such a postulated immunopathology to work, SAg expression by an infectious virus is the *conditio sine qua non.* To date, there is only one example of SAg expression by a human-pathogenic virus: the nucleocapsid (NC) of rabies virus. The superantigenicity of rabies virus NC, suggested by the work of Scott-Algara et al.,[24] seems to exhibit some special features, such as the induction of rather low proliferative responses of T cells from normal donors (even in the presence of exogenous IL-2), or the induction of Vß-specific unresponsiveness instead of Vß-specific activation in T cells from HIV-infected individuals.[25] The possible implications of rabies virus-encoded superantigenicity for human autoimmune diseases remain to be established.

At the moment, the discrepancy between the powerful action of bacterial SAgs in various experimental systems and in human diseases like the toxic shock syndrome, and the poor evidence for SAg activity in human-pathogenic viruses, has to be emphasized. The mere emphasis on a (suspected, proclaimed, or really existing) "Vß-bias" in certain human autoimmune diseases cannot substitute for a direct demonstration of SAg activity in these clinical entities. So far, the lack of evidence for SAg expression in human-pathogenic viruses (with the above-mentioned exception of rabies virus) weakens the potential impact of SAgs on autoimmunity in man.

Viral 'transactivating' proteins and autoimmunity.

Members of all three subgroups of the retroviruses – RNA-oncoviruses, lentiviruses, and spumaviruses – have been shown to encode for so-called "transactivating" regulatory proteins: HIV-1 (and HIV-2) encoded tat, HTLV-1 (and HTLV-2) encoded tax-proteins (tax I and II), equine infectious anemia virus tat, visna virus tat, bel-1 of human foamy virus, and taf of simian foamy virus type 1.

HIV tat and HTLV-1 tax are unique among viral and cellular transactivators in that they are able to activate a large set of cellular genes encoding for immune mediators, such as interleukins or growth hormone for immune cells. HTLV-1 tax seems to display the broadest activating potential, including a transcriptional activation of genes coding for IL-1, IL-2, IL-2R, IL-6, IFN-γ, GM-CSF, TNF-α, and others.[26-30] Since some of these virus-induced immune mediators are potent proinflammatory factors, it seems reasonable to assume that tax-induced cytokine production might under certain circumstances trigger inflammation (and perhaps autoimmune reactions) in virus-infected tissues. HIV-1 (either the infectious virus itself or the isolated tat antigen) has been shown to enhance, among other things, the expression of IL-4R, TNF-α, and TNF-β, and to down-modulate the expression of certain other cellular genes, e.g. the genes for the p80 form of the TNF receptor or MHC class I molecules.[31,32] A particularity of tat concerns the fact that the protein is secreted by HIV-infected cells and can be taken up by non-infected cells (as shown by Mann and Frankel[33]). This finding provides a potentially very important mechanism for HIV-induced cytokine production in *non-infected* cells.

Several autoimmune-like syndromes have been described in association with human retroviral infections. These autoimmune-like complications of retroviral diseases in man can present typical features of both systemic and organ-restricted autoimmune diseases. HTLV-1, the etiological agent of adult T-cell leukemia, induces in addition a syndrome called HTLV-1 associated myelopathy/tropical spastic paraparesis (HAM/TSP), a non-neoplastic, immunological disease of the central nervous system (CNS).[34,35] TSP is a neurological syndrome of progressing spasticity and weakness of motor functions, accompanied by disturbances of superficial and proprioceptive sensations of varying degree. The clinical and histological features show strong similarities to multiple sclerosis (MS), and indeed, TSP is a differential diagnosis to MS in endemic areas for HTLV-1. To date, autoreactive T and B cell responses to CNS autoantigens have not been systematically analyzed in patients suffering from TSP.

Experimental evidence for the transactivating potential of HTLV-1 tax stems from both a variety of cell culture experiments and a limited number of *in vivo* transgenic approaches. When transfected into glial cells, the HTLV-1 tax gene can, for example, induce the expression of MHC class I antigens in normally class I-negative astrocytes. The recent identification of glial cell-specific and HTLV-1 long-terminal-repeat (LTR)-specific DNA-protein complexes could provide the molecular basis for such a tax-mediated transactivation of cellular genes in astroglia.[36] Since a large percentage of infiltrating mononuclear cells in CNS lesions from TSP patients consists of $CD8^+$ cells, it was suggested that they might be cytotoxic against viral (and perhaps self-) antigens presented on class I-expressing glial cells. Another characteristic of TSP patients is the occurrence of oligoclonal Ig synthesis and highly elevated levels of IL-6 in the cerebrospinal fluid (CSF). This has led to speculation as to a local, IL-6 driven immunoglobulin synthesis in the brain in TSP. Indeed, tax-expressing cerebral endothelial cells were shown to express very large amounts of IL-6 constitutively,[27] thereby suggesting a mechanism of tax-triggered B cell-help by locally produced cytokines. As mentioned before, the proportion of autoantibodies in CSF immunoglobulins in TSP patients has not been evaluated.

Other non-neoplastic, immunopathological diseases related to HTLV-1 infections are HTLV-1 associated arthropathy (HAAP), systemic lupus erythematosus (SLE) and Sjögren's syndrome.[37,38] The frequent complications of inflammatory processes in the joints of HTLV-1 infected patients has given new substance to speculations concerning a viral origin of rheumatic diseases. The recent demonstration by Kitajima al.[39] of an inflammatory arthropathy resembling rheumatoid arthritis (RA) in mice transgenic for HTLV-1 tax supports such a possible link between retroviral infections and certain forms of autoimmune rheumatic diseases. Interestingly, the incidence and severity of arthritis in these tax-transgenic mice was directly correlated with the level of tax expression in the joints. Although the exact immunopathologic events leading to HAAP in tax-transgenic mice have not been outlined, a local production of IL-1 and IL-6 in tax-expressing synovial tissue was suggested as a likely explanation.[39] Production of these proinflammatory cytokines can be detected in supernatants of synovial tissue cultures from patients with HAAP. It therefore seems reasonable to assume that tax-induced cytokine production could cause and perpetuate the chronic inflammatory process seen in the joints of some HTLV-1 infected individuals.

Sjögren's syndrome (SS) has been associated with both HIV-1 and HTLV-1 infections. An association between HTLV-1 and SS has been suggested indirectly through case reports of HTLV-1 infected patients with TSP who developed features of SS. The clinical aspects of exocrinopathy in HIV-infections show striking similarities to SS, although the pattern of autoantibody-production in HIV-associated SS seems to differ from that seen in "classical" SS. Furthermore, as already described in systemic lupus erythematosus (SLE) patients, a large percentage of SS patients have antibodies reacting with HIV-1 p24.[40] However, no contribution of these antiviral antibodies to pathology has yet been demonstrated. Similar to HTLV-1-induced HAAP in transgenic mice, HTLV-1 tax-transgenic mice have been reported which develop an exocrinopathy resembling SS.[41] The exocrinopathy in these mice starts with an abnormal proliferation of ductal epithelial cells in the submandibular, parotid and minor salivary glands. In later stages, lymphocytic infiltrations are noted in the glands. As the proliferation of ductal epithelium progresses, the number of infiltrating cells increases and leads to destruction of acinus architecture and thickening of basement membranes. As already noted in HTLV-1 transgenic mice developing inflammatory arthropathy,[39] the degree of salivary gland pathology corresponds to the level of tax gene expression in the affected organ.[41]

The above-mentioned experimental evidence leaves little doubt that retroviral transactivating proteins have the potential to modify immune reactivities in various ways. Furthermore, the striking resemblance in symptomatology of certain transgenic mouse models[39,41] for human retroviruses with the autoimmune-like complications found in retrovirus-infected patients suggests that these immune abnormalities might be directly correlated to the expression of transactivating viral genes in the respective target tissue. It should be emphasized, however, that an autoimmune component (autoreactive T cell or B cell responses) has not been reported in any of the HIV or HTLV-1 transgenic animals. So far, we are therefore obliged to keep the two concepts of "autoimmunity" and "immunopathology" separate. Although the circumstantial evidence for the involvement of retroviruses as likely etiologic agents in autoimmune diseases has steadily increased during the past few years, a direct link cannot be postulated at present.

B cell "mitogenic" viral envelope proteins and autoimmunity.

Viral infections are known to be associated quite frequently with transient episodes of polyclonal B cell activation. This virus-triggered polyclonal B cell stimulation should be clearly distinguished from the isotypic bias during viral infections, leading to a predominance of IgG2a, which is not caused by particular properties of viral antigens but by IFN-γ production during infection. Although the exact mechanism(s) leading to this non-specific B cell activation by viruses is/are not known, a direct activation of lymphocytes by B cell "mitogenic" viruses has been proposed to be operative in certain viral infections. *In vitro*, a great number of rather distantly related viruses have been shown to exhibit B cell "mitogenic" properties, including members of the herpes-, rhabdo-, myxo-, paramyxo-, toga-, and retrovirus families (Table 1).[42-47] Interestingly, although the molecular mechanisms underlying this polyclonal B cell stimulatory potential of viruses seem to be rather diverse, and include, for example, both T cell-dependent and -independent forms of B cell activation, in almost all cases the "mitogenic" principle is encoded by the respective viral envelope glycoprotein.

In two cases of human-pathogenic viruses, the molecular basis for the "mitogenic" properties of viral envelope proteins has been more extensively investigated in recent years: B cell activation caused by the gp160 envelope glycoprotein of HIV-1[48] and by the hemagglutinin glycoprotein (HA) of influenza A viruses.[44,45] As will be outlined below, B lymphocyte activation in these two examples of B cell "mitogenic" antigens is achieved by two completely different molecular mechanisms, namely either by the induction of an abnormal T cell helper function in HIV-infected $CD4^+$ lymphocytes,[48-52] or by

TABLE 1 B cell "mitogenic" viruses

Virus	Family	"Mitogenic" antigen	T cell dependence	Molecular mechanism
Vesicular stomatitis virus	Rhabdoviridae	unknown	no	unknown
Sindbis virus	Alphaviridae	glycoprotein	no	unknown
Rauscher leukemia virus	Retroviridae	gp70	no	unknown
Sendai virus	Paramyxoviridae	hemagglutinin	no	unknown
Herpes simplex virus (type 1/2)	Herpesviridae	unknown	unknown	unknown
Herpes virus Saimiri	Herpesviridae	unknown	yes	membrane TNF-α; CD2/CD58 interaction
Human immunodeficiency virus	Retroviridae	gp160	yes	IL-6 production; membrane TNF-α
Influenza virus	Orthomyxoviridae	hemagglutinin	no	BCR interaction (??)

a T cell-independent, direct interaction of a "superstimulatory" viral antigen with cell surface receptors on B lymphocytes (HA).[53,54]

Polyclonal B cell activation is a rather frequently observed, early complication of HIV infection.[56-58] Individuals infected with HIV-1 have serum antibody titers to a wide range of self- and foreign antigens. Autoantibodies in HIV-infection appear early after seroconversion and persist throughout the disease course. The origin of these autoantibodies and their contribution to immunopathology are unknown, although speculations as to a pathogenic role have been made. The most frequently discussed hypothesis with regard to

HIV-induced autoantibody production is that the virus or viral proteins themselves induce a generalized polyclonal B cell activation that leads to elevated levels of all kinds of antibodies, including those reacting against self. This hypothesis is strongly supported by experimental evidence *in vitro*.

When normal, untransformed $CD4^+$ T lymphocytes are infected with HIV *in vitro*, they acquire the ability to stimulate the production of extraordinarily high amounts of IgM, IgG, and IgA by both autologous and allogeneic B lymphocytes in the absence of any further stimulation.[46,47] This implies that HIV-infected peripheral T cells can provide an excellent helper function for polyclonal Ig synthesis in the absence of cognate interaction between T and B lymphocytes. Induction of Ig synthesis in B cells depends upon a cell-to-cell contact with the infected T lymphocyte; in contrast, supernatants from the same T cells cannot mimic the helper activity.[48] The molecular mechanism for this Ag-nonspecific, MHC-unrestricted, contact-dependent polyclonal B cell activation by HIV-infected T cells has been further investigated and the accomplished data point to a crucial role for the viral envelope glycoprotein gp160.

Infectious HIV or the isolated HIV envelope protein gp160 can induce helper function for Ig-synthesis in normal T lymphocytes.[48-52] HIV-gp-160 induces a moderate up-regulation of various cell surface molecules on T cells, including HLA-DR, LFA-1α, LFA-1β, and ICAM-1. When these activated T cells are confronted with B lymphocytes, B cells are stimulated to Ig-synthesis. Physical contact between T and B cells is essential in the activation process and fixation of the gp-160-activated T cells prior to B cell encounter abrogates the response. Interestingly, addition of exogenous IL-6, but not IL-2 or IL-4, restores the B cell stimulatory capacity of paraformaldehyde-fixed T cells. Gp-160 also induces IL-6R and CD23 molecule expression on B cells when added to cultures of T and B cells; CD23 expression can only be observed in those B cells that form conjugates with T cells.[48-50] Altogether, these observations have led to the suggestion of a two-step model for B lymphocyte activation by gp-160: the first step requires T cell contact-dependent interaction of several cell surface molecules, and the second step is mediated by cytokine (IL-6) secretion.[52] This means that HIV infection induces an abnormal, enhanced T helper function in infected or gp-160-stimulated T cells that subsequently leads to Ag-nonspecific, polyclonal B cell stimulation and hyperglobulinemia, early hallmarks of HIV infection which might be the main cause of autoreactivity in later stages of the disease.

In the second case, B cell stimulation by the influenza virus HA molecule,[44,45] the underlying molecular mechanism for the "mitogenic" effect of the viral envelope protein, is an example of a T cell-independent B cell stimulation, which seems to be largely independent of soluble factors.[53,54] When resting murine B cells are confronted with inactivated influenza viruses or the purified HA envelope protein, they respond with a vigorous proliferative response and Ig synthesis. The "mitogenic" capacity of the whole virus particle shows a strict correlation with the HA-subtype of the respective virus strain, whereas the other viral proteins seem to be dispensable in the activation mode. Indeed, polyclonal B cell activation can be accomplished by the purified HA glycoprotein alone. According to their HA-subtype, the viruses have been classified into groups of high (H2, H4, H6, H12), medium (H3, H5, H8, H9), and low (H1, H7, H10, H11) B cell activators. HA-triggered B cell proliferation is T cell-independent and can be induced by either infectious or (UV-) inactivated virus, or by infected, paraformaldehyde-fixed cells in the absence of any exogenously added cytokines.[53] HA induces a remarkably strong synthesis of immunoglobulins in responding B cells, with the IgG2b, IgG3, and IgM subtypes being the predominant isotypes.[53] In addition, virus-activated B cells exhibit a hyperexpression of various activation markers, including MHC class II, IL-2R, ICAM-1, and B7 molecules, and serve as excellent antigen-presenting cells (APC) for T lymphocyte responses (O. Rott et al., unpublished data).

With regard to the molecular mechanism responsible for this particular example of a B cell "mitogenic" viral glycoprotein, evidence for a possible involvement of the B cell receptor (BCR) complex has been obtained. Virus-triggered B cell proliferation can be blocked by the simultaneous addition of $F(ab')_2$-fragments of anti-Ig monoclonal antibodies.[53] In addition, influenza viruses are able to induce a specific growth arrest in the anti-μ sensitive B cell lymphoma line WEHI-231, a widely used tumor model for tolerance induction in immature B cells. Interestingly, the "tolerogenic" capacity of HA in the WEHI-231 system correlates with the B cell stimulatory potential for mature B cells of the protein, as highly stimulatory HA-subtypes are highly "tolerogenic", whereas low stimulatory HA-subtypes are only marginally effective.[51] Finally, the signal transduction pathway of influenza virus-activated B cells has been analyzed. The results of these experiments indicate that the virus activates resting B cells by a protein kinase C (PKC)-dependent pathway, without simultaneously evoking a detectable enhancement of phosphoinositide (PI) turnover or Ca^{2+}-mobilization in responding lymphocytes.[54] Collectively, the currently available data on the activation mode of the B cell "mitogenic" influenza virus HA therefore suggest that the virus is operating as a T cell-independent, "superstimulatory" type-2 antigen by directly interacting with B cell surface receptors, most likely including the BCR itself.[53,54]

Can polyclonal B cell stimulation by the influenza virus HA *in vivo* manifest features of an autoimmune disease? Indeed, some experimental evidence for such a hypothetical pathomechanism already exists. Repeated infections with influenza viruses were shown to result in the appearance of rheumatoid factor (RF)-type autoantibodies in infected mice.[58] Even more strikingly, transgenic mice expressing a highly "mitogenic" HA subtype (H2) under the control of the insulin promotor in the pancreas are not tolerant for the transgenic product, but produce anti-HA Abs and, in later stages, develop an immune-mediated form of diabetes, characterized by T and B cell infiltrations in the pancreas and the appearance of multiple anti-islet Abs.[57] This striking example of a transgenic mouse model, where the transgenic product does not behave as a tolerogen but as an *immunogen*, argues that the prolonged, tissue-restricted expression of a B cell "mitogenic" viral antigen can induce local lymphocyte activation at the site of protein expression, finally resulting in a chronic inflammatory process that presents typical features of an organ-specific autoimmune disease.

Concluding remarks: lessons from viruses

Viruses have been implicated in the pathogenesis of almost every kind of autoimmune disease. Nonetheless, the precise mechanisms involved have never been elucidated to satisfaction. In recent years, a number of new and exciting concepts on virus-encoded molecular mechanisms for immunopathology have been described. With some of these new concepts in mind, transgenic mouse models have been created which bear, in some instances, a remarkably strong resemblance to human (virus-induced?) "autoimmune" diseases, such as rheumatoid arthritis, diabetes mellitus, or connective tissue diseases (SLE and Sjögren's syndrome). Common to all of these rather divergent examples of virus-mediated immunopathology is the finding that the isolated expression of a single viral antigen can lead to pathological alterations which are histologically and clinically nearly indistinguishable from the corresponding human autoimmune disease. Work in the future will certainly further help to close the gap between our knowledge on virus-triggered immunopathology and the still rather dissatisfying concepts as to the molecular mechanisms underlying human autoimmune diseases.

Summary

Viruses are commonly held to be the most important etiologic and physiopathologic factors in autoimmune diseases. However, despite the long-standing implication of these agents in the induction of autoreactive immune responses, the exact molecular mechanisms responsible for virus-mediated breaking of natural tolerance still remain elusive. In recent years, a number of new physiopathologic mechanisms by which viruses are able to interfere with the normal functioning of the immune system have come to the forefront of immunological research. These new immunopathologic principles comprise as diverse mechanisms as T cell 'superantigenic' proteins, retrovirus-encoded 'transactivating' regulatory proteins, or B cell 'mitogenic' viral envelope proteins. In the present paper, the possible implications of these novel examples for virus/immune cell-interactions for our understanding of how viruses might trigger autoimmune diseases have been critically evaluated.

References

1 Fleischer B, Schrezenmeier H. T-cell stimulation by staphylococcal enterotoxins: clonally variable response and requirement for major histocompatibility complex class II molecules on accessory or target cells. J Exp Med 1988;167:1697.
2 Fleischer B, editor. Biological significance of superantigens. Basel: Karger, 1992.
3 Marrack P, Kappler J. The staphylococcal enterotoxins and their relatives. Science 1990;248:705.
4 Fleischer B, Sjögren HO, editors. Superantigens. Heidelberg: Springer Verlag,

1991.
5 Acha-Orbea H, Held W, Waanders GA, Shakhov AN, Scarpellino L, Lees RK, MacDonald HR. Exogenous and endogenous mouse mammary tumor virus superantigens. Immunol Rev 1993;131:5.
6 Rammensee H-G, Kroschewski R, Frangoulis B. Clonal anergy induced in mature Vß6$^+$ T lymphocytes on immunizing Mls-1^b mice with Mls-1^a expressing cells. Nature 1989;339:541.
7 Kawabe Y, Ochi A. Selective anergy of Vß8$^+$, CD4$^+$ T cells in staphlococcus enterotoxin B-primed mice. J Exp Med 1990;172:1065.
8 Rellahan BL, Jones LA, Kruisbeek AM, Fry AM, Matis LA. In vivo induction of anergy in peripheral Vß8$^+$ T cells by staphylococcal enterotoxin B. J Exp Med 1990;172:1091.
9 Rott O, Fleischer B. A superantigen as virulence factor in an acute bacterial infection. J Infect Dis 1994;169:1142.
10 Woodland DL, Happ MP, Gollob KJ, Palmer E. An endogenous retrovirus mediating deletion of αβ T cells? Nature 1991;349:529.
11 Acha-Orbea H, Shakhov AN, Scarpellino L, Kolb E, Müller V, Vessaz-Shaw A, Fuchs R, Blöchlinger K, Rollini P, Billotte J, Sarafidou M, MacDonald HR, Diggelmann H. Clonal deletion of Vß14 bearing T cells in mice transgenic for mammary tumour virus. Nature 1991;350:207.
12 O'Rourke AM, Mescher MF, Webb SR. Activation of polyphosphoinositide hydrolysis in T cells by H-2 alloantigen but not Mls determinants. Science 1990;249:171.
13 Liu H, Lampe MA, Iregui MV, Cantor H. Conventional antigen and superantigen may be coupled to distinct and cooperative T-cell activation pathways. Proc Natl Acad Sci USA 1991;88:8705.
14 Oyaizu N, Chirmule N, Yagura H, Pahwa R, Good RA, Pahwa S. Superantigen staphylococcal enterotoxin B-induced T-helper cell activation is independent of CD4 molecules and phosphatidylinositol hydrolysis. Proc Natl Acad Sci USA 1992;89:8035.
15 Rott O, Wekerle H, Fleischer B. Protection from experimental allergic encephalomyelitis by application of a bacterial superantigen. Int Immunol 1992;4:347.
16 Gaur A, Fathman CG, Steinman L, Brocke S. SEB induced anergy: Modulation of immune response to T cell determinants of myoglobin and myelin basic protein. J Immunol 1993;150:3062.
17 Kalman B, Lublin FD, Lattime E, Joseph J, Knobler RL. Effects of staphylococcal enterotoxin B on T cell receptor Vß utilization and clinical manifestations of experimental allergic encephalomyelitis. J Neuroimmunol 1993;45:83.
18 Bette M, Schäfer MK-H, Van Rooijen N, Weihe E, Fleischer B. Distribution and kinetics of superantigen-induced cytokine gene expression in mouse spleen. J Exp Med 1993;178:1531.
19 Marrack P, Blackman M, Kushnir E, Kappler J. The toxicity of staphylococcal enterotoxin B in mice is mediated by T cells. J Exp Med 1990;171:455.
20 Fischer H, Dohlsten M, Andersson U, Hedlund G, Ericsson P, Hansson J, Sjögren HO. Production of TNF-alpha and TNF-beta by staphylococcal enterotoxin A activated human T cells. J Immunol 1990;144:4663.
21 Schiffenbauer J, Johnson HM, Butfiloski EJ, Wegrzyn L, Soos JM. Staphylo-

coccal enterotoxins can reactivate experimental allergic encephalomyelitis. Proc Natl Acad Sci USA 1993;90:8543.

22 Brocke S, Gaur A, Piercy C, Gautam A, Gijbels K, Fathman CG, Steinman L. Induction of relapsing paralysis in experimental autoimmune encephalomyelitis by bacterial superantigen. Nature 1993;365:642.

23 Rott O, Mignon-Godefroy K, Fleischer B, Charreire J, Cash E. Superantigens induce primary T cell responses to soluble autoantigens by a non-Vß specific mechanism of bystander activation. Cell Immunol 1995;161:158.

24 Lafon M, Lafage M, Martinez-Arends A, Ramirez R, Vuillier F, Charron D, Lotteau V, Scott-Algara D. Evidence for a viral superantigen in humans. Nature 1992;358:507.

25 Scott-Algara D, Lafon M, Vuillier F, Pialoux G, Dauguet C, Dighiero G. Viral superantigen-induced hyporesponsiveness of T cells and polyclonal B cell activation in HIV-1 infection. Eur J Immunol 1994;24:2595.

26 Yodoi J, Uchiyama T. Diseases associated with HTLV-I: virus, IL-2 receptor dysregulation and redox regulation. Immunol Today 1992;13:405.

27 Rott O, Tontsch U, Fleischer B, Cash E. Interleukin-6 production in "normal" and HTLV-1 tax-expressing brain-specific endothelial cells. Eur J Immunol 1993;23:1987.

28 Yamashita I, Katamine S, Moriuchi R, Nakamura Y, Miyamoto T, Eguchi K, Nagataki S. Trans activation of the human interleukin-6 gene by human T-lymphotropic virus type 1 Tax protein. Blood 1994;84:1573.

29 Yoshida M. Mechanism of transcriptional activation of viral and cellular genes by oncogenic protein of HTLV-1. Leukemia 1994;8(Suppl 1):S51.

30 Albrecht H, Shakhov AN, Jongeneel CV. Trans activation of the tumor necrosis factor alpha promoter by the human T-cell leukemia virus type I Tax1 protein. J Virol 1992;66:6191.

31 Westendorp MO, Li-Weber M, Frank RW, Krammer PH. Human immunodeficiency virus type I Tat upregulates interleukin-2 secretion in activated T cells. J Virol 1994;68:4177.

32 Zauli G, Furlini G, Re MC, Milani D, Capitani S, La Placa M. Human immunodeficiency virus type 1 (HIV-1) tat-protein stimulates the production of interleukin-6 (IL-6) by peripheral blood monocytes. New Microbiol 1993;16:115.

33 Mann DA, Frankel AD. Endocytosis and targeting of exogenous HIV-1 Tat protein. EMBO J 1991;10:1733.

34 Gessain A, Barin F, Vernant JC, Gout O, Maurs L, Calender A, De The G. Antibodies to human T-lymphotropic virus type-I in patients with tropical spastic paraparesis. Lancet 1985;2:407.

35 Osame M, Matsumoto M, Usuku K, Izumo S, Ijichi N, Amitani H, Tara M, Igata A.Chronic progressive myelopathy associated with elevated antibodies to human T-lymphotropic virus type I and adult T-cell leukemia-like cells. Ann Neurol 1987;21:117.

36 Tillmann M, Wessner R, Wigdahl B. Identification of human T-cell lymphotropic virus type I 21-base pair repeat-specific and glial cell-specific DNA-protein complexes. J Virol 1994;68:4597.

37 Kitajima I, Maruyama I, Maruyama Y, Ijichi S, Eiraku N, Mimura Y, Osame

M. Polyarthritis in human T lymphotropic virus type I associated myelopathy. Arthr Rheum 1989;32:1342.

38 Olsen RG, Tarr MJ, Mathes LE, Whisler R, Du Plessis D, Schulz EJ, et al. Serological and virological evidence of human T-lymphotropic virus in Systemic Lupus Erythematosus. Med Microbiol Immunol 1987;176:53.

39 Iwakura Y, Tosu M, Yoshida E, Takiguchi M, Sato K, Kitajima I, Nishioka K, Yamamoto K, Takeda T, Hatanaka M, Yamamoto H, Sekiguchi T. Induction of inflammatory arthropathy resembling rheumatoid arthritis in mice transgenic for HTLV-1. Science 1991;253:1026.

40 Talal N, Garry RF, Schur PH, et al. A conserved idiotype and antibodies to retroviral proteins in systemic lupus erythematosus. J Clin Invest 1990;85:1866.

41 Green JE, Hinrichs SH, Vogel J, Jay G. Exocrinopathy resembling Sjögren's syndrome in HTLV-1 tax transgenic mice. Nature 1989;341:72.

42 Goodman-Snitkoff G, Mannino RJ, McSharry JJ. The glycoprotein isolated from vesicular stomatitis virus is mitogenic for mouse B lymphocytes. J Exp Med 1981;153:1489.

43 Del Prete G, De Carli M, D'Elios MM, Müller Fleckenstein I, Fickenscher H, Fleckenstein B, Almerigogna F, Romagnani S. Polyclonal B cell activation induced by herpesvirus saimiri-transformed human $CD4^+$ T cell clones. J Immunol 1994;153:4872.

44 Scalzo AA, Anders EM. Influenza viruses as lymphocyte mitogens. I. B cell mitogenesis by influenza A viruses of the H2 and H6 subtypes is controlled by the I-E/C subregion of the major histocompatibility complex. J Immunol 1985;134:757.

45 Scalzo AA, Anders EM. Influenza viruses as lymphocyte mitogens. II. Role of I-E molecules in B cell mitogenesis by influenza A viruses of the H2 and H6 subtypes. J Immunol 1985;135:3524.

46 Pahwa S, Pahwa R, Saxinger C, Gallo RC, Good RA. Influence of the human T lymphotropic virus/lymphadenopathy associated virus on human lymphocytes: Evidence for immunosuppressive effects and polyclonal B cell activation by banded viral preparations. Proc Natl Acad Sci USA 1985;82:8198.

47 Schnittman SM, Lane HC, Higgins SE, Folks T, Fauci AS. Direct polyclonal activation of human B lymphocytes by the acquired immune deficiency syndrome virus. Science 1986;233:1084.

48. Chirmule N, Kalyanaraman VS, Lederman S, Oyaizu N, Yagura H, Yellin MJ, Chess L, Pahwa S. HIV-gp160-induced T cell-dependent B cell differentiation. J Immunol 1993;150:2478.

49 Chirmule N, Kalyanaraman VS, Saxinger C, Wong-Staal F, Ghrayeb J, Pahwa S. Localization of the B-cell stimulatory activity of HIV-1 to the carboxyl terminus of gp41. AIDS Res Hum Retroviruses 1990;6:299.

50 Macchia D, Parronchi P, Piccinni MP, Simonelli C, Mazzetti M, Ravina A, Milo D, Maggi E, Romagnani S. In vitro infection with HIV enables human CD4+ T cell clones to induce noncognate contact-dependent polyclonal B cell activation. J Immunol 1991;146:3413.

51 Delfraissy J-F, Wallon C, Boué F, Barré-Sinoussi F, Galanaud P. Tumor necrosis factor alpha inhibits the competence signal delivered by HIV to normal B cells. J Immunol 1991;146:1516.

52 Boué F, Wallon C, Goujard C, Barré-Sinoussi F, Galanaud P, Delfraissy J-F. HIV induces IL-6 production by human B lymphocytes. Role of IL-4. J Immunol 1992;148:3761.

53 Rott O, Cash E. Influenza virus hemagglutinin induces differentiation of mature resting B cells and growth arrest of immature WEHI-231 lymphoma cells. J Immunol 1994;152:5381.

54 Rott O, Charreire J, Semichon M, Bismuth G, Cash E. B cell superstimulatory influenza virus (H2-subtype) induces B cell proliferation by a PKC-activating, Ca^{2+}-independent mechanism. J Immunol 1995;154:2092.

55 Lane HC, Masur H, Edgar LC, Wahlen G, Rook AH, Fauci AS. Abnormalities of B-cell activation and immunoregulation in patients with the acquired immunodeficiency syndrome. N Engl J Med 1983;309:453.

56 Muller S, Richalet P, Laurent-Crawford A, Barakat S, Riviere Y, Porrot F, Chamaret S, Briand JP, Montagnier L, Hovanessian A. Autoantibodies typical of non-organ-specific autoimmune diseases in HIV-seropositive patients. AIDS 1992;6:933. Comment in AIDS 1993;7:893.

57 Ditzel HJ, Barbas SM, Barbas III CF, Burton DR. The nature of the autoimmune antibody repertoire in human immunodeficiency virus type I infection. Proc Natl Acad Sci USA 1994;91:3710.

58 Fazekas G, Rajnavölgyi E, Kurucz I, Sintar E, Kiss K, Laszlo G, Gergely J. Isolation and characterization of IgG2a-reactive autoantibodies from influenza virus-infected BALB/c mice. Eur J Immunol 1990;20:2719.

59 Roman LM, Simons LF, Hammer RE, Sambrook JF, Gething M-JH. The expression of influenza virus hemagglutinin in the pancreatic ß cells of transgenic mice results in autoimmune diabetes. Cell 1990;61:383.

HANS-GEORG RAMMENSEE

Physiological function of MHC molecules

MHC stands for Major Histocompatibility Complex and designates a gene region originally discovered as coding for antigens responsible for allograft rejection. MHC genes have been found in all vertebrates analyzed in this respect; those investigated most intensively are the MHCs of the mouse, also called H-2 (for Histocompatibility gene number 2) and the human MHC, also called HLA, for Human Leukocyte Antigen.

The physiological function of MHC genes is, of course, not to code for antigens recognized by allograft-rejecting lymphocytes. Rather, MHC molecules play a central role in regulating the specific immune system.[1] Their exact function has been discovered stepwise over the last 20 years and can be summarized in one sentence: MHC molecules are peptide receptors that collect peptide samples inside the cell, transport them to the cell surface and present them there to T lymphocytes for immune surveillance. These peptides can be derived from any protein synthesized in the cell, as is the case for MHC class I molecules, or from proteins which are endogenous or exogenous to the cell (e.g., after phagocytosis), as is the case for MHC class II molecules. On the whole, a cell expresses in the range of 10^5 MHC copies of a given MHC molecule on its surface; these 10^5 MHC molecules present 10^3-10^4 different peptides at individual copy numbers of 10^0-10^4. Thus, at the cell surface, a sample of roughly every cellular protein is displayed by MHC molecules. Individual combinations of MHC molecules and peptides can be recognized by the T cell receptor (TCR) of T lymphocytes, thus making it possible for the immune system to "look inside the cell", that is, to control the protein composition of the cell – a task not suitable for antibodies. T cells are usually tolerant to those combinations of peptides and MHC molecules that have always been around, that is, to the normal self. As soon as new proteins occur in the cell, for example after virus infection, peptides derived from the new proteins will also be presented by MHC molecules at the cell surface. The new combinations of peptides and MHC molecules so far unknown to the T cells will be considered as foreign and attacked, with the possible result of destruction of the virus-infected cells by cytotoxic T lymphocytes.

MHC genes are extremely polymorphic. They are encoded on chromosome 17 of the mouse and on the human chromosome 6 (Fig. 1). The principal human class I genes are HLA-A, B, and C, coding for the heavy chains of class I molecules; the light chain, also known as $ß_2$-microglobulin, is coded on chromosome 15. Both chains associate non-covalently to build the molecule depicted in Fig. 2 as a ribbon diagram. The structure, as determined by X-ray crystallography by Don Wiley's group,[2] impressively shows the peptide receptor function. The α_1 and α_2 domains of the heavy chain form a peptide accommodation device made up of two α-helices as borders of a groove and a ß-pleated sheet as its floor.

A number of years ago, we decided to try to isolate the–then only hypothetical–peptides associated with MHC class I molecules. The approach developed by Olaf Rötzschke and Kirsten Falk,[3-7] at that time undergraduate students in my laboratory, is outlined in Fig. 3: MHC molecules are immunoprecipitated from a large number of

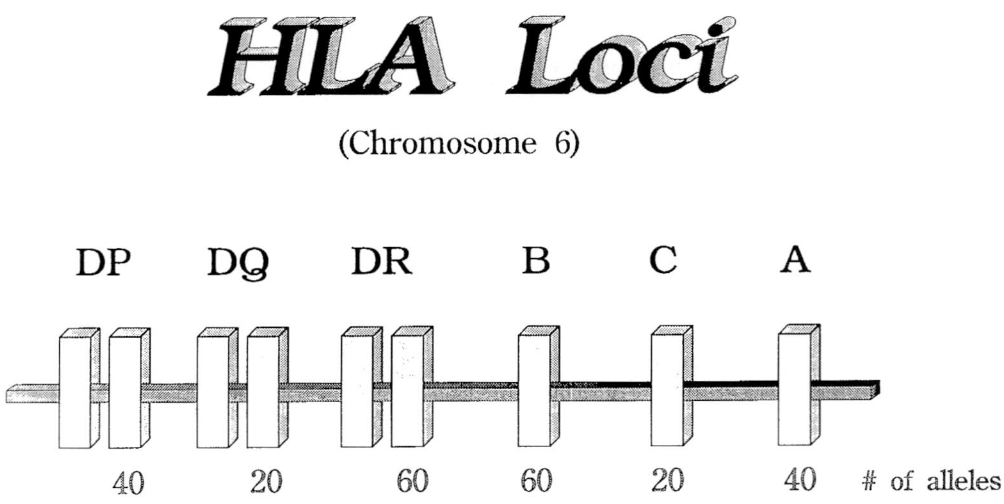

FIGURE 1 The principal HLA loci on chromosome 6.

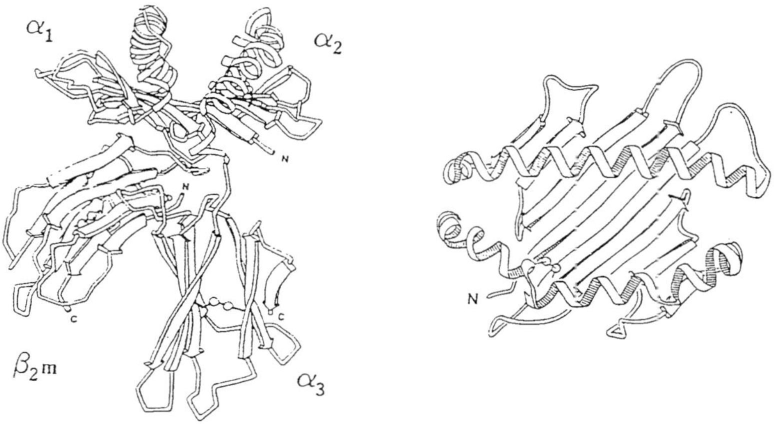

FIGURE 2 Left ribbon model of a human HLA class I molecule; right, top view into the peptide binding groove. (Adapted from Bjcrkman et al., Nature 1987;329:506–12.)

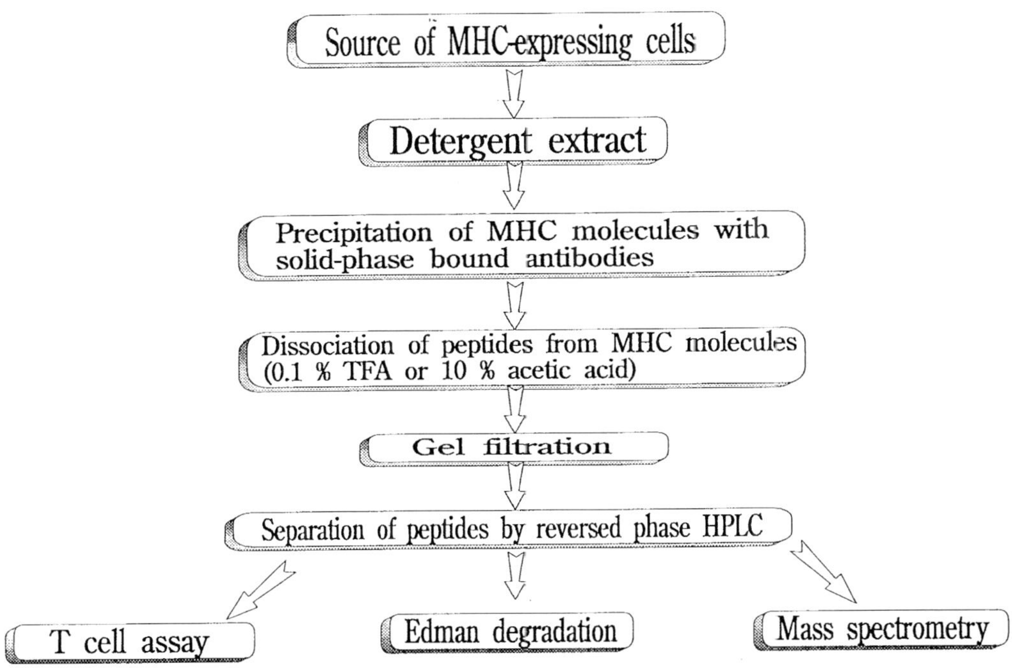

FIGURE 3 The technical approach for isolation of MHC-associated peptides.

detergent-lysed cells. Peptides are then dissociated from MHC molecules by acid treatment, and are separated by reversed phase HPLC. The resulting HPLC fractions can then be analyzed by testing them for recognition by T cells, by Edman degradation, or by mass spectrometry. In the following, I will concentrate on peptide sequence analysis by Edman degradation. Fig. 4 shows a typical HPLC profile of peptides eluted from an MHC class I molecule, in this case, HLA-A2. The individual HPLC peaks can be collected separately and sequenced separately; this results in identification of individual peptides associated with the MHC molecule in question. If we do something very unusual, however, we can obtain a different type of information. Combining all individual HPLC fractions and sequencing them as a very complex mixture of peptides, an approach we call pool sequencing, reveals the common features shared between the thousands of peptides associated with a single MHC species. For HLA-A2, for example, these common features are a peptide length of 9 amino acids, a conserved occupancy of position

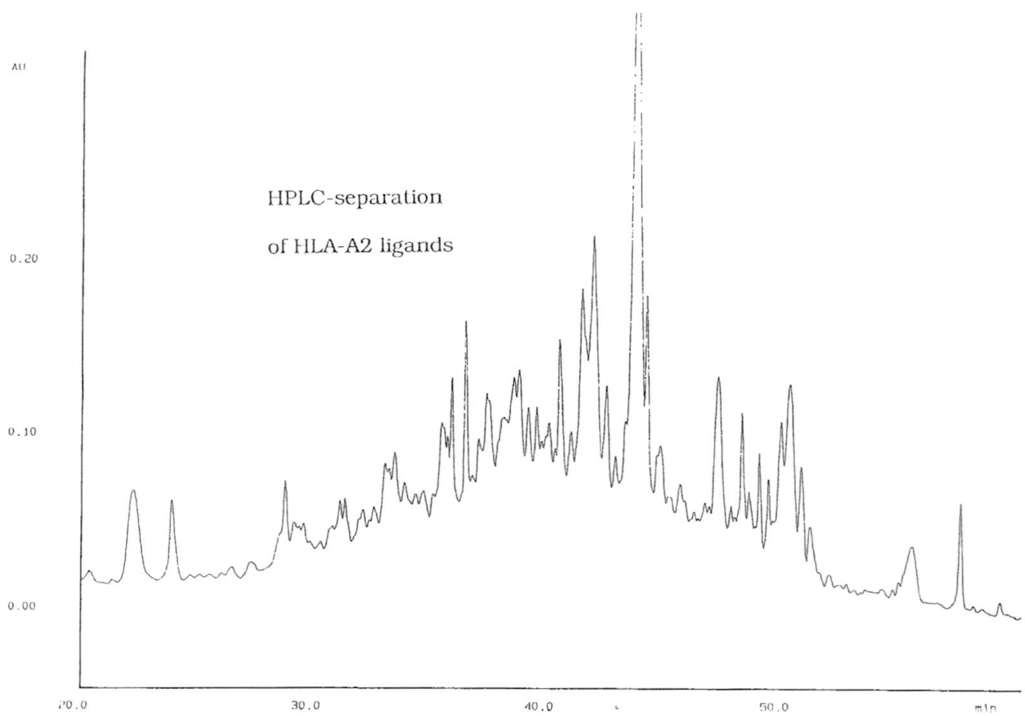

FIGURE 4 A typical reversed-phase HPLC profile of HLA class I peptide ligands.

2 of the peptides with either Leu or Val, and a conserved occupancy of position 9 with Val or Leu, as pictured in the cartoon in Fig. 5. We called these conserved positions at 2 and 9 "anchors", assuming that they are anchored in complementary pockets of the MHC molecule. This pattern, or motif, confirmed by other groups afterwards,

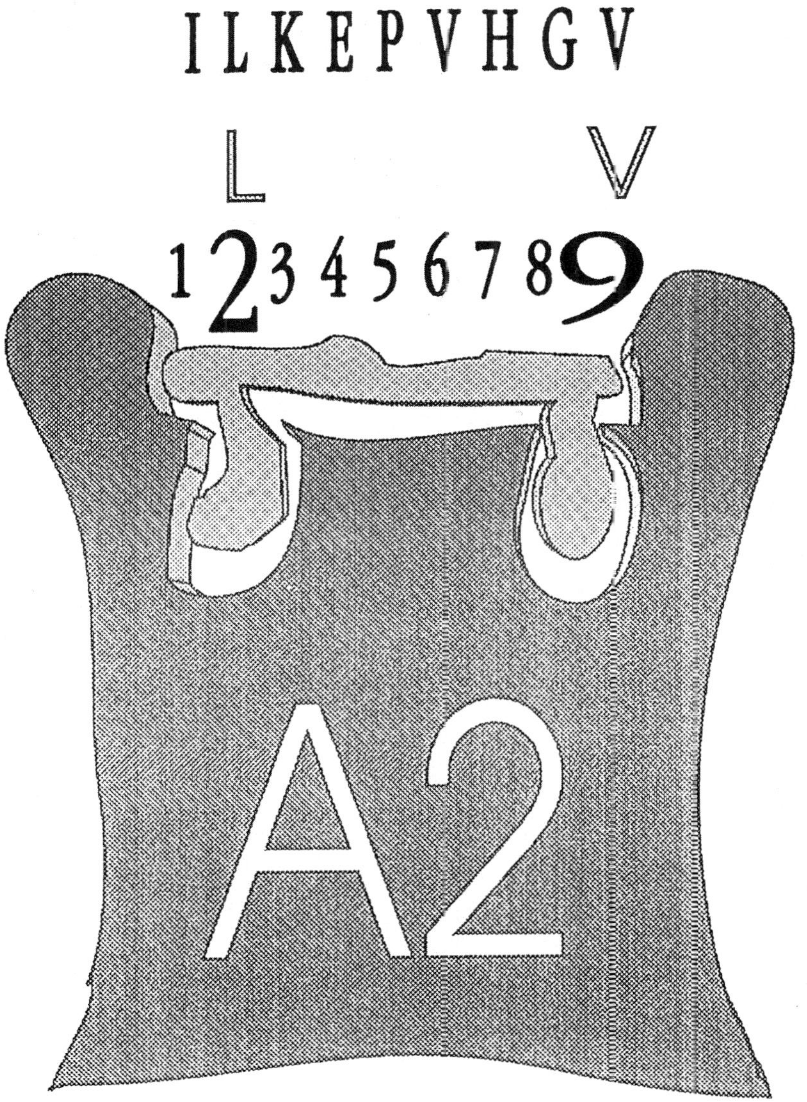

FIGURE 5 A naive cartoon illustrating peptide-MHC class I interaction and specificity for the example of HLA-A2 and an HIV-derived peptide.

represents the individual peptide receptor specificity of HLA-A2 molecules, since other MHC class I molecules have different motifs, as we had shown first for the mouse molecules H-2K^d, D^b and K^b, and later on for a large number of HLA molecules.

Fig. 6 shows in detail the combined information we can get from both pool sequencing and sequencing of individual ligands, with the example of HLA-A1: The motif requires 9-mers with negative charges at position 3, and a Tyr at the C-terminal position 9. In addition, there are preferences at non-anchor positions, like positions 2 and 6. The individual ligands generally adhere to the motif, although there are exceptions, for example regarding peptide length. The protein origin of the individual peptides illustrates that proteins located in all cellular compartments can donate peptides for MHC

HLA-A1

Position	1	2	**3**	4	5	6	7	8	**9**
Anchors		T	**D**				L		**Y**
		S	**E**						
Preferred		L		P	G	G			
				G	N	V			
				I	Y	I			

	Sequence	Protein
Ligands	A T **D** F K F A M **Y**	Cyclin-like protein
	I A **D** M G H L K **Y**	Proliferation cell nuclear antigen
	M I **E** P R T L Q **Y**	Ribosomal protein S16
	Y T S D Y F I S **Y**	Ets-1
	L T **D** P G V L D **Y**	Unknown
	V S **D** I V G P D G L V **Y**	Fibrillarin
	Y T **D** Y G G L I F N S **Y**	Cytochrome C oxidase II
	Q S **E** D G S H T I Q T M **Y**	HLA class I α chain
	Y L **D** D P D L K **Y**	Cytosine methyl transferase
	S T **D** H I P I L **Y**	Fructose-6-amino transferase
	D S **D** G S F F L **Y**	IgG4 279-287
	G T **D** E X R N X **Y**	Unknown
	V S **D** P Y N X K **Y**	Unknown
	V A **D** K V H X M **Y**	Unknown
	Y T A V V P L V **Y**	J-chain 102-110
	Y T N P Q F N V **Y**	Unknown
	E T X X P D W S **Y**	Unknown
	F T **D** V N S X X R Y	Unknown
T-cell epitopes	E A **D** P T G H S **Y**	MAGE1 161-169
	V S **D** G G P N L **Y**	Influenza A PB1 591-599

FIGURE 6 Detailed information on the HLA-A1 motif, ligands, and T cell epitopes.

Basic HLA-A motifs

	Position 1	2	3	4	5	6	7	8	9		
A*0101		*T,S*	**D,E**						**Y**		
A*0201		**L,M**				*V*			**V,L**		
A*0205		*V,L*				*I,V*			**L**		
A3		*L,V*	*F,Y*			*I,M*	*I,L*		**K,Y**		
A*1101		*V,I*	*M,L*				*L,I*		**K**	**K**	**K**
A24		**Y**			*I,V*	*F*			**I,L,F**		
A*3101		*L,V*	*F,L*			*L,F*			**R**		
A*3302			*A,I*						**R**		
A68.1		**V,T**							**R,K**		

FIGURE 7 The basic features of the HLA-A motifs known to date.

ligands; there are nuclear, cytosolic, or mitochondrial proteins, for example.

Figs. 7 and 8 show lists of the basic HLA-A and B motifs known

Basic HLA-B motifs

	Position 1	2	3	4	5	6	7	8	9
B7		**P**	*R*						**L,F**
B8			**K**		**K R**				**L**
B*2702		**R**							**F,Y,I**
B*2705		**R**							**L,F,R**
B*3501		**P**							**Y,F,M**
B*3701		**D,E**			*V,I*				**F,M,L**
B*3801		*H*	*D,E*						**F,L**
B*39011		**R,H**				*I,V*			**L**
B*3902		**K,Q**			*I,L,F*				**L**
B40		**E**	*F,I*						**L**
B*4402		**E**							**F,Y**
B*4403		**E**							**Y,F**
B*5101		**A,P,G**							**F,I**
B*5102		**P,A,G**	*Y*						**I,V**
B*5103		**A,P,G**	*Y*						**V,I,F**
B*5201		*Q*	*F,Y*		*L,I*			*I,V*	**I,V**
B53		**P**							
B*5801		**A,S,T**		*P,E,K*	*V,I,L*				**F,W**
B60 (B*40012)		**E**					*I,V*		**L**
B61 (B*4006)		**E**	*F,I*			*I*			**V**
B62 (B*1501)		**Q,L**			*I,V*				**F,Y**
B*7801		**P,A,G**				*I,L*			

FIGURE 8 The basic features of the HLA-B motifs known to date.

to date; basic means that only the most important features have been included, since many of the details have been omitted for brevity. These lists illustrate the common features of the motifs: Predetermined length, two anchors, one always at the C-terminus and either hydrophobic or charged. An obvious consequence of this functional polymorphism is that each individual recognizes different parts of a foreign protein, as shown in Fig. 9 for the example of influenza-virus nucleoprotein. An immediate practical consequence for vaccine design is that a single peptide cannot be a useful vaccine for an entire population; as a minimum, a future peptide vaccine must contain a mixture of peptides fitting to a number of the more frequent HLA allelic products in order to cover a high percentage of the population.

The motifs shown in Figs. 7 and 8 suggest that a number of different HLA molecules have rather similar motifs. However, this does not necessarily mean that such molecules select the same peptides from a given protein, since HLA molecules with the same basic motifs might differ considerably in their preferences at non-anchor

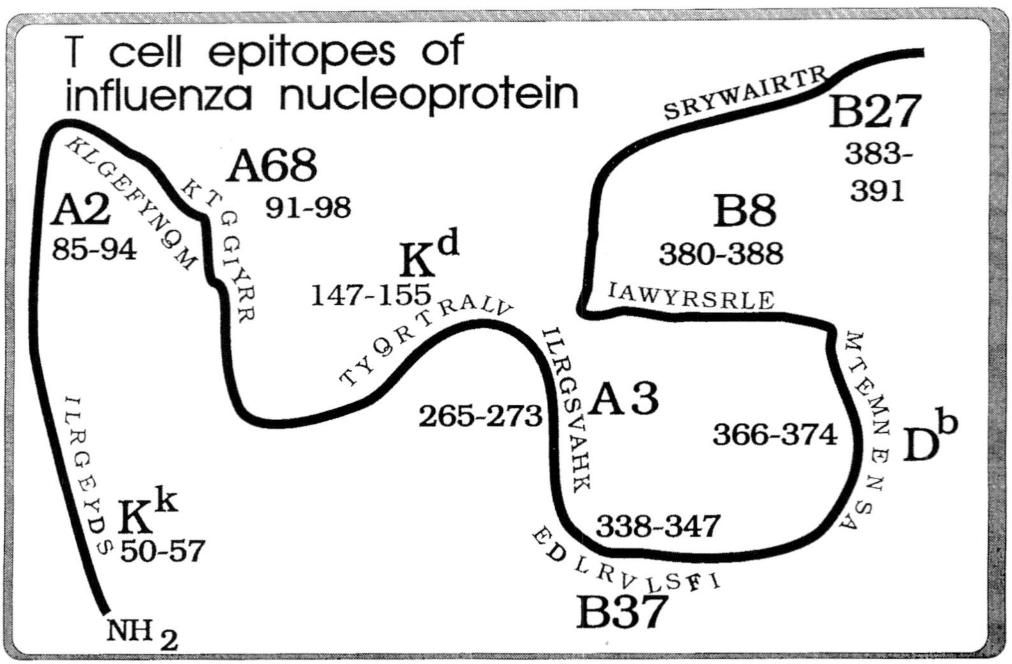

FIGURE 9 MHC class I restricted T cell epitopes within a viral protein reflect the functional MHC polymorphism.

positions; this is shown for one example, the case of B*5101 and B*5102, in Fig. 10. The only difference between the two is in one amino acid, position 171 of the heavy chain, predicted to be involved in the contact site for the first amino acid in the peptide ligand, at P1. The basic motifs are very similar, requiring a small amino acid as one anchor at position 2 and a hydrophobic one at the C-terminal P9. P1 of the B*5101 motif, however, has a strong preference for negatively charged residues, based as expected on a positive counter-charge at the predicted P1 contact site on the MHC molecule (Fig. 11). B*5102, however, does not have a positive charge there and thus does not have a charge preference for ligands at P1; this molecule tends to prefer hydrophobic residues at P1. In addition to showing that HLA molecules with very similar basic motifs might still differ considerably in their peptide receptor fine specificity, this example can be used to address two additional important points. First, the fine structure of the MHC groove and the peptide receptor fine specificity can be impressively correlated so that one can explain the other, thanks to the detailed crystallographic analysis

HLA-B*5101 *(Behcet +)*	Position 1 **2** 3 4 5 6 7 8 **9**	
Anchors	**A** F **P** I **G**	
Ligands	Y **P** F K P P K **V**	UBC5, yeast 61-68
	D **A** H I Y L N H **I**	Thymidylate synthase 253-261
	T **G** Y L N T V T **V**	GBLP 192-199
	D **A** Y A L N H T **L**	No match
	L **P** P E V N R Q **I**	No match
HLA-B*5102 *(Behcet -)*	1 **2** 3 4 5 6 7 8 **9**	
Anchors	**P** Y I **A** V **G**	
Ligands	Y **A** Y D G K D Y **I**	MHC I α chain 140-148
	Y **P** F K P P K **V**	UBC5, yeast 61-68
	L **P** P G R I I K X	No match
	L **P** F T V I L **v**	CDC25 homol. 560-567
	T **G** Y L N T V T **V**	GBLP 192-199
	F **A** Y D G K D Y **I**	MHC I α chain
	F **P** S E I V G K R **I**	Ribosomal protein
	M **P** W F K G W K **V**	EF1a 208-216

FIGURE 10 A comparison of B*5101 and B*5102 motifs.

Residues contributing to the P1 contact site

	63	66	163	167	171	P1 ligand residues
HLA-B*5101 *(Behcet +)*	ASN	ILE	LEU	TRP	HIS	Negative charge; hydrophobic
HLA-B*5102 *(Behcet -)*	ASN	ILE	LEU	TRP	TYR	Hydrophobic

FIGURE 11 Residues contributing to the P1 contact site in B*5101 and B*5102 molecules.

of the MHC structure. Similar considerations are also valid for the interactions of anchors with their respective pockets (see Figs. 12 and 13). Second, the observed differences in the peptide receptor fine specificity of the closely related HLA molecules considered correlate with a differential association with an autoimmune disease. B*5101 is highly associated with Behçet's disease, whereas B*5102 is not. Thus, one can draw conclusions about features of the putative peptide recognized by putative autoreactive T cells–such a peptide should fit to the associated but not to the non-associated HLA molecule. It should follow the general B51 basic motif but have a negative charge at P1. Similar considerations of peptide receptor specificity with disease association might also be informative for other closely related HLA molecules with differential disease associations, as will be discussed later on for some class II alleles.

So far, I have covered the peptide receptor specificity and the functional polymorphism of HLA class I molecules, as well as some

Residues contributing to the P2-accommodating pocket

HLA	9	24	45	63	66	67	P2 specificity
A*0201	Phe	Ala	Met	**Glu**	Lys	Val	aliphatic
B*2705	His	Thr	**Glu**	**Glu**	Ile	Cys	positive
B*3501	Tyr	Ala	Thr	Asn	Ile	Phe	P
B*5103	Tyr	Ala	Thr	Asn	Ile	Phe	P,A,G
B*60 (B*40012)	His	Thr	Lys	**Glu**	Ile	Ser	negative

FIGURE 12 Relation between pocket 2 specificity and the structure of the P2-accommodating pocket.

Motif groups (MHC class I)

Anchor at P2

P2	C-terminus (P8,**P9**,P10,P11)	Examples
aliphatic	aliphatic	A2
pos.	pos. or hydrophobic	B*2705
hydrophobic	pos.	A3,A11,A31,A33,B*3901
P	hydrophobic	B7,B35,B53,H-2L^{d}
P,A,G	hydrophobic	B*5101,B*5102,B*7301
Y	hydrophobic	H-2K^{d},A24
neg.	hydrophobic	H-2K^{k},B37,B40,B60

FIGURE 13 Motif groups of MHC class I.

consequences of those features. In the next part, I will address the way that peptides are generated, or processed, in the cell. If we look again at Fig. 9, the question arises as to how a cell manages to produce all the peptides fitting to the MHC molecules co-expressed in the cell. The processing machinery is probably rather conserved, so that one has to assume that it is able to produce all the peptide ligands for all MHC molecules expressed in a species and beyond. For example, a cell from an individual with HLA-A2 will be able to produce all the peptides for HLA-A1, A3, A11, A24, A31, A33 and so forth, as indeed indicated by transfection experiments. On the other hand, earlier work from my laboratory has shown that a cell contains only peptides fitting to the MHC molecules co-expressed in that cell, and not peptides fitting to other MHC molecules. For example, in a total peptide extract of a cell expressing A1 but not A2, we will not find peptides fitting to A2. These considerations as well as a number of additional results from a number of laboratories, most notably that of Pramod Srivastava, have led us to our current working hypothesis on peptide generation for class I loading, as depicted in Fig. 14. Briefly, peptides longer than the final class I ligands are generated in the cytosol and transported into the ER lumen through the Transporter Associated with Antigen Processing (TAP). In the ER lumen, these peptides are chaperoned by gp96

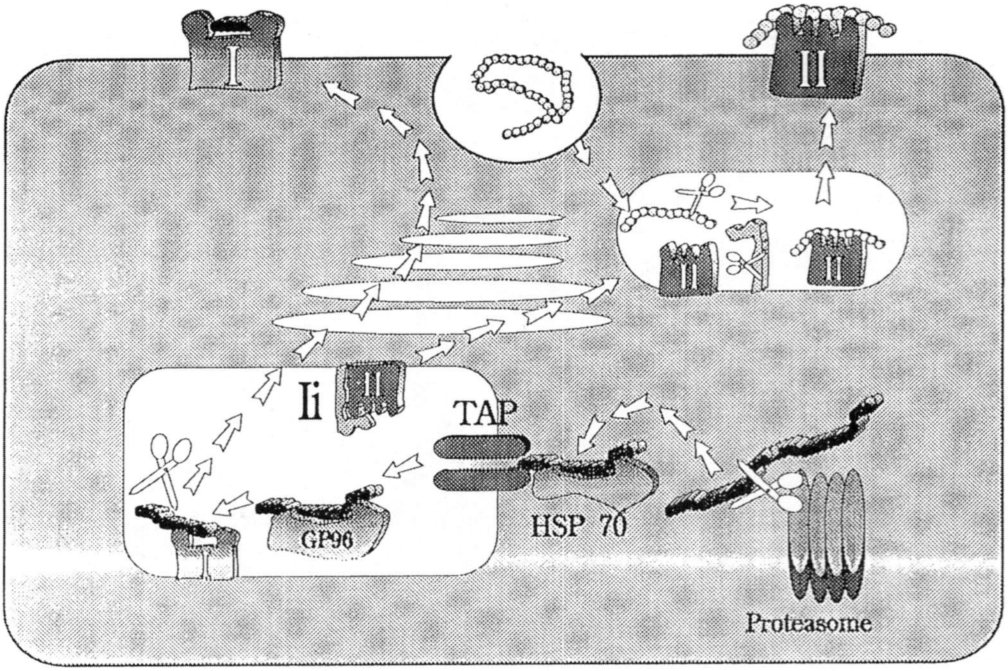

FIGURE 14 A simplified and partially hypothetical view of antigen processing.

molecules and transferred from there to MHC molecules. During these events, and possibly as late as after first contact with the MHC molecules, the peptides are trimmed until the correct length is reached. This is, to reemphasize the point, our working hypothesis and not proven fact. In order to test the hypothesis, we attempted to isolate gp96 molecules and to analyze the bound peptides. The technical approach is shown in Fig. 15. The results were rather surprising. A very complex mixture of peptides was detected after TFA treatment of gp96. Pool sequencing of these peptides indicated a very clear peptide motif (Fig. 16). Quite a number of individual ligands could be sequenced in addition; most of them fit very well to the hydrophobic/hydrophilic motif (Fig. 17). Unexpectedly, however, many of these peptides were derived from gp96 itself. Inspection of the cleavage site generating the detected peptides indicated that the cleavage was almost always between two hydrophobic residues (Fig. 18). We are now investigating the possibility that gp96 is a protease, in addition to being a chaperone. If so, we would have to modify our hypothesis on class I restricted antigen process-

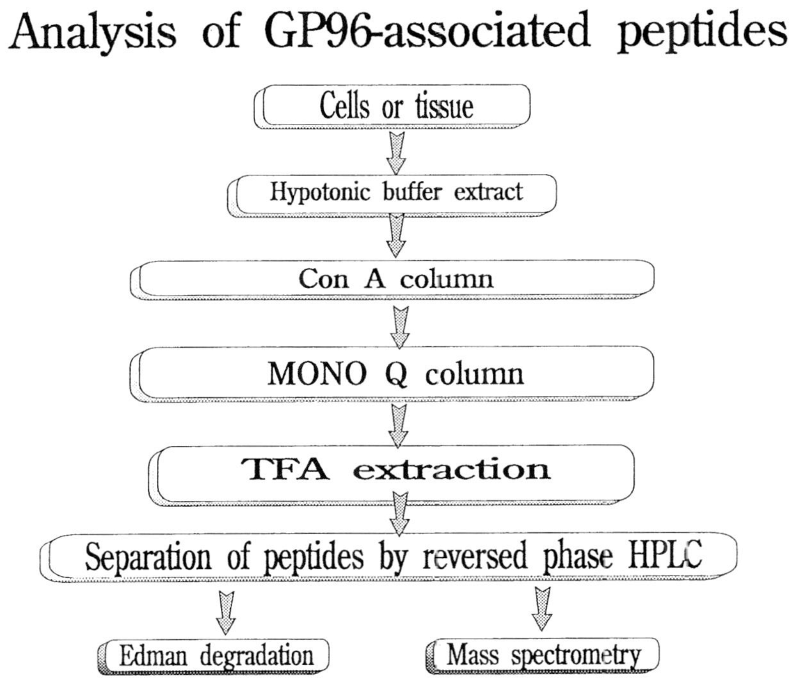

FIGURE 15 Technical approach to the analysis of gp96-associated peptides.

ing (Fig. 19): The long peptides coming into the ER lumen would be bound and cleaved by gp96, according to the ligand/cleavage

A putative gp96 ligand motif

Position

	1	2	3	4	5	6	7	.....	....
hydrophobic	V,I, L,F M,Y A								
hydrophilic		E,D K,Q R,N T,S							
Proline				P	P	P			

FIGURE 16 A putative gp96 ligand motif.

Putative natural ligands of gp96 molecules

	Position											
	1	**2**	3	4	5	6	7	.....	....			
Conserved residues	**V,I, .L,F M,Y A**	**E,D K,Q R,N T,S**		P	P	P						
Source:												
unknown (m)	**Y**	**D**	R	V	R	P	D	G	K	S	K	
unknown (m)	**L**	**E**	L	L	X	S	K	R	R	K	.	.
G19P (m)	**I**	**E**	E	W	K	T	A	R	E	E	K	Q
G19P (m)	**I**	**E**	L	G	T	H	P	E	L	D	T	G
CAP 7 (b)	**F**	**R**	N	W	I	D	S	I	N	Q	P	P
Calreticulin	**A**	V	L	G	L	D	L					
Pyr. dehydr.	X	**H**	F	Q	D	E	X	E	V	E	.	.
Hemogl. (b)	**A**	**S**	H	L	P	S	D	F	T	P	.	.
Hemogl. (b)	H	L	P	S	D	F	T	P	A	V	.	.
Hemogl. (b)	**L**	V	V	Y	P	W	T	Q	R	F		
GP96 (m)	**I**	**Q**	L	D	G	L	N	A	S	Q	I	R
GP96 (m)	R	**E**	L	I	S	N	A	S	D	A	L	.
GP96 (m)	**I**	**N**	F	P	I	Y	V	.	.	.		
GP96 (m)	**M**	**N**	D	I	K	P	I	W	Q	R	P	S

FIGURE 17 A list of putative natural peptide ligands of gp96 molecules.

motif, and then transferred to MHC while further trimming should occur until the required size for high-affinity binding is reached.

We must therefore learn a great deal more about the role of gp96

Cleavage sites of putative natural gp96 ligands

(between hydrophobic residues)

G19P (m)	S	**V**	/	**I**	**E**	L
CAP 7 (b)	V	**A**	/	**F**	**R**	N
Calreticulin b	S	**F**	/	**A**	V	L
Hemogl. (b)	T	**L**	/	**A**	**S**	H
Hemogl. (b)	A	S	/	H	L	P
Hemogl. (b)	R	**L**	/	**L**	V	V
GP96 (m)	E	**A**	/	**I**	**Q**	L
GP96 (m)	F	**L**	/	R	**E**	L
GP96 (m)	Q	**F**	/	**I**	**N**	F
GP96 (m)	E	**L**	/	**M**	**N**	D
GP96 (m)	T	**F**	/	**W**	**K**	E

FIGURE 18 Cleavage sites of putative natural gp96 ligands.

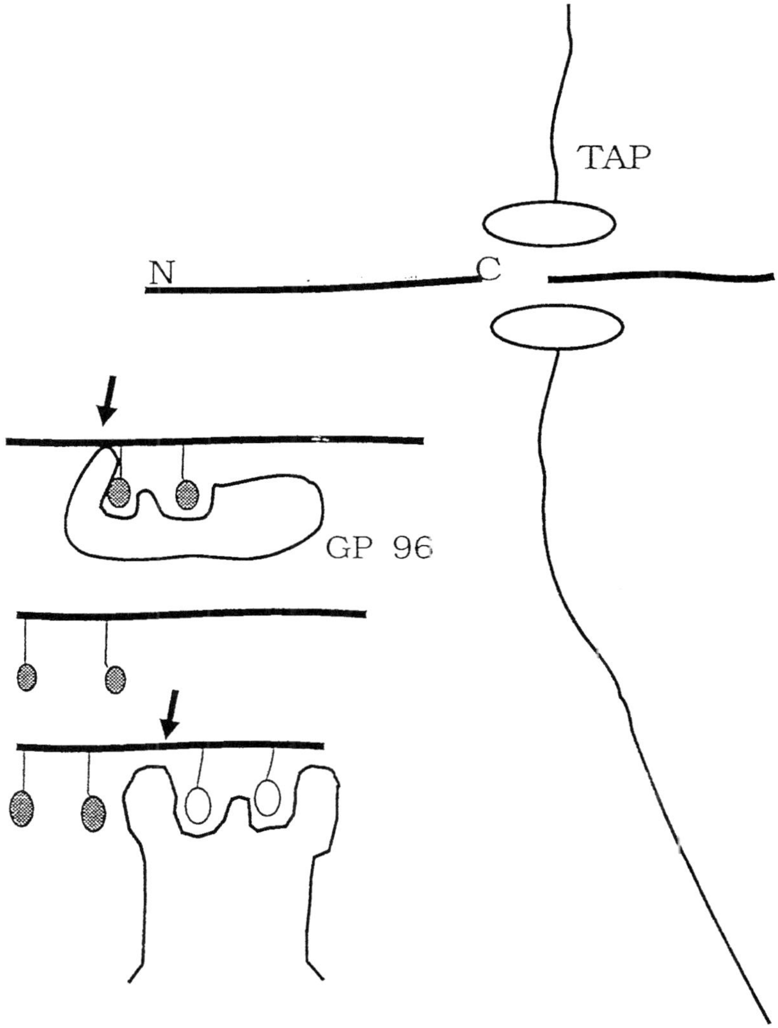

FIGURE 19 A modified hypothesis on gp96 involvement in processing. (See text for explanation.)

in antigen processing in particular, and about the whole process in general.

To conclude this article, I will turn to MHC class II. Fig. 20 shows a comparison of the peptide receptor function of both MHC classes. Class II ligands are also anchored in specific pockets, as determined recently by X-ray analysis of a monopeptidic DR1 crystal by Don Wiley's group. The peptides, however, are longer than class

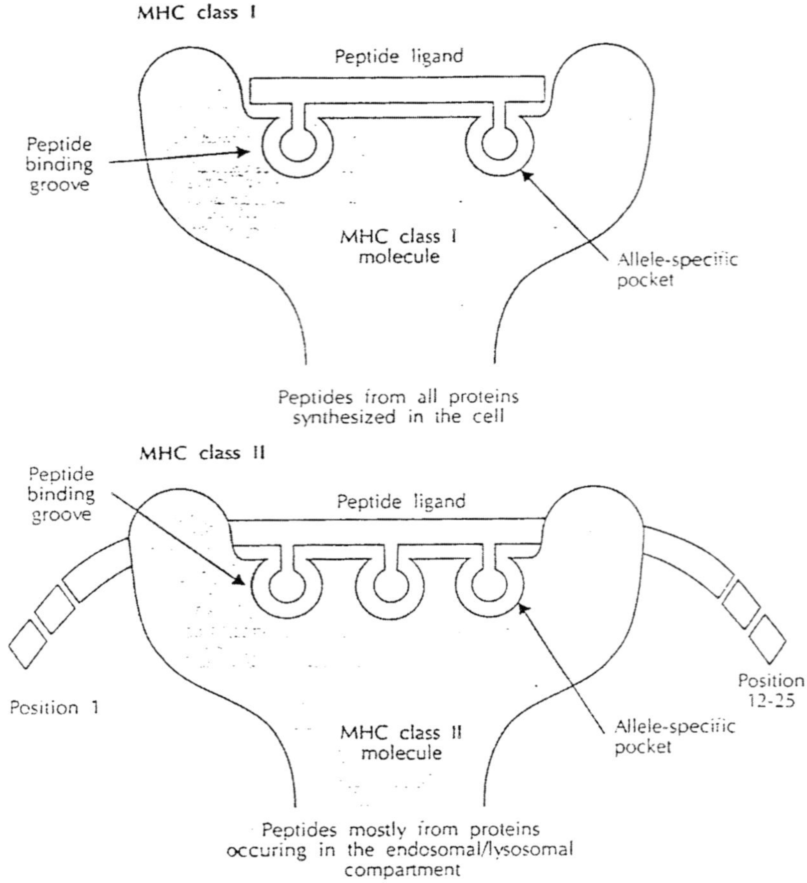

FIGURE 20 A comparison of the peptide receptor functions of both MHC classes. (Adapted from Rammensee et al.[7].)

I ligands; the extra length is accommodated by protrusion of both the N- and C-terminal parts out of the groove's ends.

To analyze the peptide ligands and the peptide receptor specificity of MHC class II molecules, we used the same approach as for class I molecules (see Fig. 3). Since the distance between the N-termini of ligands and the first anchor varies for class II ligands, pool sequencing was more difficult to interpret than for class I. Anchor residues, however, could be clearly observed in the form of clusters spanning three cycles of Edman degradation (Fig. 21), whereby the first cluster, representing the anchor for P1, spans cycles 4, 5 and 6.

HLA-DR4Dw4 pool sequencing

Cycle of Edman degradation	Amino acid residues (in pmol)*																		
	I	V	L	F	W	M	A	G	T	S	Y	P	H	Q	N	E	D	K	R
1	139.4	290.4	217.6	162.6	17.3	39.7	166.3	-	-	120.0	114.7	29.7	31.3	27.5	27.4	125.4	173.3	96.3	35.6
2	64.2	160.5	133.9	29.5	7.6	18.7	69.9	-	-	-	32.8	**155.8**	-	-	7.2	44.4	-	94.2	-
3	63.7	111.7	136.9	**68.2**	8.4	**28.4**	**121.1**	-	-	70.6	**80.6**	121.7	35.6	98.7	**44.8**	**122.7**	221.7	**151.3**	69.0
4	58.7	96.1	107.0	**205.0**	**14.9**	**30.8**	103.8	-	-	55.6	**143.2**	96.7	30.8	70.8	**56.1**	111.6	191.3	129.2	88.2
5	75.1	**140.4**	97.5	160.6	**18.6**	**33.8**	102.5	-	-	37.7	**151.3**	66.6	23.7	**144.3**	49.2	111.2	169.6	72.4	**101.1**
6	**98.4**	120.8	94.1	187.1	**19.7**	30.5	118.1	-	-	44.9	101.0	59.8	21.7	65.0	46.4	83.7	127.8	**110.3**	**117.4**
7	87.0	**168.7**	101.9	107.4	14.4	20.7	**147.7**	-	-	33.3	55.7	49.1	15.1	52.8	45.6	96.2	179.2	**123.9**	100.6
8	51.7	92.9	85.7	62.6	10.3	20.1	**162.3**	-	-	33.1	35.0	50.1	11.9	75.2	41.0	96.2	**191.3**	80.6	**154.2**
9	52.9	96.1	72.1	**117.2**	8.2	25.9	139.5	396.6	61.3	43.8	26.0	51.0	13.8	50.5	**61.2**	58.5	166.9	56.7	98.7
10	37.3	78.0	76.5	41.0	6.3	19.7	116.7	220.3	60.1	46.7	23.1	50.7	**17.7**	38.3	**75.8**	51.8	231.3	42.5	66.7
11	34.6	55.2	69.0	23.6	3.7	23.7	170.3	121.4	40.2	54.6	24.4	57.5	**21.0**	51.3	46.6	46.7	126.9	42.3	58.5
12	42.7	41.4	70.1	17.4	2.7	16.9	160.7	90.4	28.8	38.3	20.3	40.6	14.8	**63.5**	27.8	43.9	134.2	37.7	72.3
13	33.6	33.6	53.8	13.7	1.9	21.9	172.8	78.0	22.0	33.3	17.4	29.6	11.1	**76.2**	27.1	35.8	67.6	50.0	59.6
14	30.8	29.5	42.8	18.6	2.2	12.4	139.6	57.0	25.1	25.9	16.5	27.3	9.9	58.4	25.5	35.5	49.7	**52.1**	51.7
15	21.7	23.1	29.8	18.2	1.6	8.6	65.0	44.7	23.9	35.1	16.3	29.1	9.0	43.0	18.3	28.0	43.3	36.8	51.1
16	13.9	16.7	20.7	12.6	1.9	**13.3**	37.6	34.6	16.8	19.1	12.0	26.6	7.4	59.3	14.2	27.3	34.6	26.4	41.4
17	9.3	14.6	15.8	7.8	12	9.8	27.6	27.8	11.8	11.1	7.3	21.4	6.0	32.1	10.1	22.6	23.6	17.4	**63.8**
18	6.6	11.2	12.1	5.7	1.0	**27.1**	20.7	25.4	9.2	9.4	7.1	16.9	5.5	18.5	7.9	15.9	20.3	13.2	42.6

FIGURE 21 An example of pool sequencing of HLA-class II associated peptides, for the example of HLA-DR4Dw4.

Thus, the average distance from the N-terminus of peptides to the

HLA-DR4Dw4 motif and ligands

	1	2	3	4	5	6	7	8	9		Source	Fraction§
Anchor or preferred residues*	**F,Y,W, I,L, V,M**			F,W,I, L, V,A, D,E **no R,K**		N,S T,Q H,R	polar, charged, aliphatic		polar, aliphatic K			
	1	2	3	**4**	5	**6**	**7**	8	**9**			
VDDTQ	**F**	V	R	**F**	D	**S**	**D**	A	**A**	SQRMEP...	HLA-A2 (28-?)	53, 56
DTQ	**F**	V	R	**F**	D	**S**	**D**	A	**A**	SQR	" (30-44)	40
	F	V	R	**F**	D	**S**	**D**	A	**A**	SQR	" (33-44)	33
	F	V	R	**F**	D	**S**	**D**	A	**A**	SQRMEP	" (33-47)	42, 49
VDDTQ	**F**	V	R	**F**	D	**S**	**D**	A	**A**	SQRM	" (28-45	46, 55
	F	V	R	**F**	D	**S**	**D**	A	**A**	SQRM	" (33-45)	45
VDDTQ	**F**	V	R	**F**	D	**S**	**D**	A	**A**	SPRGEP...	HLA-C (28-?)	50, 54
DGKD	**Y**	I	A	**L**	N	E	**D**	L	**S**	S	HLA-B44 (143-156)	
LSS	**W**	T	A	**A**	D	**T**	**A**	A	**Q**	ITQ	" (154-168)	48
LSS	**W**	T	A	**A**	D	**T**	**A**	A	**Q**	IT	" (154-167)	49
IY	**F**	R	N	Q	K	G	**S**	H	**S**	GLQPTGFL	HLA-DR4β (252-270)	61
DVA	**F**	V	K	**D**	Q	**T**	**V**	I	**Q**	NTD	bovine transferrin (68-82)	40
YDHN	**F**	V	K	**A**	I	**N**	**A**	I	**Q**	KSW	cathepsin C (170-185)	73, 55
KHKV	**Y**	A	C	**E**	V	**T**	**H**	Q	**G**	...	Igκ chain C region (80-?)	
HKV	**Y**	A	C	**E**	V	**T**	**H**	Q	**G**	L...	Igκ chain (region 81-?)	
DGP	**F**	R	I	**I**	T	V	**P**	A	**A**	LDY	unknown	86, 87
TGN	**Y**	R	I	**E**	S	V	**L**	S	**S**		Sphingolipid activator protein 3 (165-176)	
GERA	**M**	T	K	**D**	N	**N**	**L**	L	**G**	...	HSC 70 (445-?)‡	
XXX	**Y**	E	X	**A**	L	**S**	**L**	P	**S**	K...	no match	
GSLF	**V**	Y	N	**I**	T	**T**	**N**	K	**Y**	KAFLKQ	VLA-4 (229-247)	(Chicz et al., 1993)
SPEDF	**V**	Y	Q	**F**	K	G	**M**	C	**Y**	F	HLA-DQβ 3.2 chain (24-38)	(Chicz et al., 1993)
AAPYEKEVP	**L**	S	A	**L**	T	**N**	**I**	L	**S**	AQL	PAI-1 (261-281)	(Chicz et al., 1993)
GVYF	**Y**	L	Q	**W**	G	**R**	S	T	**L**	VSVS	Ig heavy chain (121-?)	(Chicz et al., 1993)
AEALERM	**F**	L	S	**F**	P	**T**	**T**	K	**T**		bovine hemoglobin (26-41)	(Chicz et al., 1993)

FIGURE 22 Motif and peptide ligands of HLA-DR4Dw4.

first anchor is 5 amino acids. Fig. 22 shows a summary of the information we obtain from sequencing class II ligands, with the example of DRB1*0401. First, the peptide motif, determined by pool sequencing, structure analysis, and ligand alignment, and defined by 5 anchors with a 9-amino acid stretch in the peptide core numbered P1 through P9 for relative peptide position (the first anchor is defined as P1). Second, a list of individual ligands, almost all of which fit well to the motif. A comparison of the motifs of several DR4 subtypes and of DR1 indicates that each class II molecule has its own allele-specific ligand motif (Fig. 23), as has also been found for cor-

Relations between DR pocket composition, preference for ligand residues, and association with rheumatoid arthritis

DR molecule	Pocket for relative position					RA association
	P1	P4	P6	P7	P9	
DRB1*0101 (DR1)						intermediate
polymorphic residues	*86 Gly*	*13 Phe, 70 Gln, 71 Arg*	*11 Leu, 13 Phe*	*28 Gln, 67 Leu, 71 Arg*	*9 Trp, 57 Asp*	
ligand specificity	**Y, V, L, F, I,** A, M, W	**L, A,** I, V, M, N, Q	**A, G,** S, T, P		L, A, I, V, N, F, Y	
DRB1*0401 (DR4Dw4)						strong
polymorphic residues	*86 Gly*	*13 His, 70 Gln, 71 Lys*	*11 Val, 13 His*	*28 Asp, 67 Leu, 71 Lys*	*9 Glu, 57 Asp*	
ligand specificity	**F, Y, W,** I, L, V, M	F, W, I, L, V, A, D, E; **no R, K**	N, S, T, Q, H, R	polar, charged, aliphatic	polar, aliphatic, K	
DRB1*0402 (DR4Dw10)						absent
polymorphic residues	*86 Val*	*13 His, 70 Asp, 71 Glu*	*11 Val, 13 His*	*28 Asp, 67 Ile, 71 Glu*	*9 Glu, 57 Asp*	
ligand specificity	**V, I, L, M**	Y, F, W, I, L, M, H, R, N; **no D, E**	N, Q, S, T, K	R, K, H, N, Q, P; rare D, E	polar, aliphatic, H	
DRB1*0404 (DR4Dw14)						strong
polymorphic residues	*86 Val*	*13 His, 70 Gln, 71 Arg*	*11 Val, 13 His*	*28 Asp, 67 Leu, 71 Arg*	*9 Glu, 57 Asp*	
ligand specificity	**V, I, L, M**	F, Y, W, I, L, V, M, A, D, E; **no R, K**	N, S, T, Q, K, D	polar, charged, aliphatic	polar, aliphatic, K	
DRB1*0405 (DR4Dw15)						intermediate
polymorphic residues	*86 Gly*	*13 His, 70 Gln, 71 Arg*	*11 Val, 13 His*	*28 Asp, 67 Leu, 71 Arg*	*9 Glu, 57 Ser*	
ligand specificity	**F, Y,** W, V, I, L, M	V, I, L, M, F, D, E	N, S, T, Q, K, D	polar, charged, aliphatic	**D, E,** Q	

FIGURE 23 A comparison of DR1 and several DR4 motifs, with consideration of pocket structure and specificity and of the association between rheumatoid arthritis (RA) and DR alleles.

responding peptide-binding motifs, e.g. by F. Sinigaglia's laboratory. In addition, Fig. 23 also shows that the predicted pocket structure is complementary to the anchor specificity. If we look, for example, at the P1 anchor, it is evident that a Gly at residue 86 of the ß-chain is associated with a preference for large aromatic residues at P1. This can be well explained by the non-existent side chain of Gly. On the other hand, if ß-chain residue 86 is a Val, the P1 preference is for the smaller aliphatic side chains, which is now explained by a reduction of the pocket space by the side chain of Val. The comparison of the ligand motifs of closely related DR4 subtypes allows an interesting consideration with regard to autoimmune disease. These alleles are differentially associated with rheumatoid arthritis (RhA): DRB1*0401 and DRB1*0404 are strongly associated with RhA, whereas DRB1*0402 is not. The P4 pocket specificity of the two associated molecules excludes positively charged residues like Arg and Lys, whereas P4 of the non-associated molecules tolerates positive side chains but excludes the negative ones. Thus, under the assumption that RhA is induced by a T cell response directed against a peptide presented by DRB1*0401 and DRB1*0404 but not DRB1*0402, one can now attempt to predict features of such a peptide: At P4, such a peptide should have a negative charge. At P1, it should have an aliphatic residue to fit to both types of P1 specificity.

Thus, similarly as for the two B51 subtypes discussed earlier, comparison of the peptide motifs of closely related but differentially disease-associated HLA molecules should allow further insight into the reasons for HLA/disease association.

In summary, we have seen that HLA molecules are peptide receptors with a peculiar specificity, and with a central regulatory function within the immune system. Detailed understanding of HLA function and specificity will contribute to increasing our insight into the workings of the immune system, and will enhance our ability to manipulate the immune responses involved in infections, malignant, and autoimmune diseases.

Note added in proof: In the meantime, we have found that our gp96 preparations are contaminated with a proteolytic enzyme, cathepsin E. Thus, the 'putative ligand motif of gp96' is in reality a cathepsin E cleavage motif (D. Arnold et al., manuscript submitted for publication).

Acknowledgement

The following individuals contributed essentially to the work covered in this article: Kirsten Falk, Olaf Rötzschke, Stefan Stevanović, Wieland Keilholz, Günther Jung, Stefan Faath, Danièle Arnold, Hansjörg Schild, Tilman Dumrese, and Thomas Friede.

References

1 Klein J. Natural history of the major histocompatibility complex. New York: J. Wiley & Sons, 1986.
2 Stern LJ, Wiley DC. Antigenic peptide binding by class I and class II histocompatibility proteins. Structure 1994;2:245-51.
3 Falk K, Rötzschke O, Stevanovic S, Jung G, Rammensee HG. Allele-specific motifs revealed by sequencing of self-peptides eluted from MHC molecules. Nature 1991;351:290-6.
4 Falk K, Rötzschke O, Takiguchi M, Gnau V, Stevanovic S, Jung G, Rammensee HG. Peptide motifs of HLA-B51, -B52, and -B78 molecules and implications for Behçet's disease. Int Immunol 1995;7:223-8.
5 Rammensee HG, Falk K, Rötzschke O. Peptides naturally presented by MHC class I molecules. Annu Rev Immunol 1993;11:213-44.
6 Rammensee HG, Friede T, Stevanovic S. MHC ligands and peptide motifs. First listing. Immunogenetics 1995;41:178-228.
7 Rammensee HG, Falk K, Rötzschke O. MHC molecules as peptide receptors. Curr Opin Immunol 1993;5:35-44.

ELISABETH MÄRKER-HERMANN

Induction of autoimmunity by bacteria

Introduction

In has now become clear that some of the most important environmental factors that may be associated with autoimmunity are infectious agents. In patients with a certain genetic predisposition, bacteria, viruses and parasites may participate in the breakdown of tolerance and the development of autoimmune disease. In this review, we will focus on the role of bacteria in the induction and maintenance of autoimmune disease. According to the classification into humoral or cellular autoimmune human diseases, the role of molecular mimicry, cross-reactive antibodies, bacterial endotoxins, bacterial and self heat shock proteins, bacteria-specific $CD4^+$, $CD8^+$ and $\gamma\delta$-TCR^+ T lymphocytes, and bacterial superantigens are discussed. Independently of the mechanisms described here, one has to keep in mind that the presence of even high titers of cross-reacting autoantibodies or of self-reacting lymphocytes is not by itself sufficient to induce clinical autoimmune disease.

Molecular mimicry and cross-reactive antibodies

The paradigmatic example of bacteria-induced autoimmune disease and antigenic mimicry is rheumatic fever, a disease which occurs shortly after pharyngeal group A streptococcal infection. The inflamed myocardium of rheumatic carditis contains deposits of antibodies and complement. There is strong evidence that the immunogens provoking that lesion may be cross-reacting streptococcal antigens. Among the candidate antigens, streptococcal M proteins, the protective surface antigens of the bacteria, contain epitopes that cross-react with major constituents of the myocardium. The extended α-helical coil structure of M proteins is rare among bacterial surface molecules but is common in mammalian proteins. M proteins share significant amino acid sequence homology with myosin heavy chain and type I keratins (M5, M6, M12, M24 proteins) or with human α-tropomyosin (M5, M6, M24) (for a review, see ref. 1). Antistreptococcal antibodies cross-recognizing those human proteins are thought to be involved in poststreptococcal autoimmune disease, namely acute rheumatic fever, rheumatic heart disease, and acute poststreptococcal glomerulonephritis. Such cross-reactive antibodies, which can be detected by means of indirect immunofluorescence, appear to be quite common in patients with rheumatic heart disease. Some of these autoantibodies can be absorbed by certain streptococcal strains, whereas others react specifically with cardiolipin and tropomyosin.[2] In Chagas disease – which is caused by the protozoan parasite *Trypanosoma cruzi* – heart autoantibodies react with laminin, while Chagasic cardiomyopathy may be due to recognition of the calcium-sequestering ATPase in the sarcoplasmic reticulum.[2]

Antigenic cross-reactivity between host and bacteria can also be encountered in acute or chronic pulmonary tuberculosis. The diversity of autoantibodies in the sera of patients infected with mycobacteria extends to rheumatoid factors, DNA and other nuclear components, thyroglobulin, spermatozoa, erythrocytes, and cerebral glycolipids, although none of these patients ever exhibit signs or symptoms suggestive of connective tissue disorders (for a review, see ref. 3).

In HLA-B27-associated arthritis and spondylarthropathy (reactive arthritis, Reiter's disease, arthritis in chronic inflammatory bowel disease, ankylosing spondylitis), several lines of evidence have suggested that bacterial products are involved in the pathogenesis of the spondylarthropathies by molecular mimicry mechanisms. In these rheumatic diseases, the molecular mimicry theory states that HLA-B27 has antigenic structural epitopes which are shared by arthritogenic gram-negative bacteria. The immune response against the bacteria would lead to an autoimmune response to HLA-B27. The existence of such shared amino acid sequences has been demonstrated for *Klebsiella pneumoniae* nitrogenase and HLA-B27,[4] as well as for other enterobacteria and the first hypervariable region of HLA-B27.[5,6] However, as reported by several more critical investigators, there is only little correlation of such serologic cross-reactivity with disease.[7] Recently, the shared epitope hypothesis has been replaced more and more by the "arthritogenic peptide model" of HLA-B27-associated spondylarthropathy (see below). Furthermore, it has now become clear that peptides derived from HLA class I molecules make up a sizable proportion of the self-peptides isolated from HLA class II molecules.[8] Thus, a third hypothesis on the role of HLA-B27 in disease pathogenesis proposes that a B27-derived peptide presented by an HLA class II molecule might cross-react with a class II-presented bacterial peptide.

Other examples of antigenic mimicry between host and bacteria are cross-recognition of blood group substances and bacterial polysaccharides, of kidney tissues and *E.coli* lipopolysaccharide, *Yersinia enterocolitica* protein and TSH receptor in Grave´s disease, as well as of brain components and the capsular polysaccharide of *Neisseria meningitidis* group B and *E.coli* K1.

Bacterial endotoxins and autoimmunity by CD5+ B cells

A distinct lineage of human B cells expressing the T-cell antigen, CD5, is considered to be involved in autoimmune phenomena. Much evidence suggests that most of the probably limited repertoire of specificities of the antibodies produced by CD5+ B cells is directed against the polar head groups of phospholipids, which form the basic matrix of all cell membranes. Such cell surface antigens (including the common cell membrane phospholipid, phosphatidylcholine) and bacterial cell surface antigens are presented to the immune system on the surfaces of phospholipid bilayers (cells, bacteria, lioposomes) which also contain endotoxin, LPS or lipid A. Insertion of endotoxins or their lipid A-containing fragments is postulated to trigger autoimmune reactions against the surface antigens of such cells by CD5+ B cells. Naturally occurring autoimmune antiphospholipid antibodies which are found in normal rabbit and human sera and antiphospholipid antibodies produced by immunization with liposomes containing lipid A belong to the IgM class.[9]

Heat shock proteins (Hsp)

Environmental stress to microbes results in major changes in their physiologic behavior. A major component of these adaptive changes is the production of heat shock proteins (hsp) that make possible the survival and continued pathogenicity of these organisms. Likewise, in the corresponding stressful situation in the mammalian system, mammalian (self) hsp60 is known to be up-regulated as a physiologic response to stimuli such as bacterial infection or rheumatic inflammation, and has been shown to be elevated in inflamed synovia of patients with rheumatoid arthritis[10] and other rheumatic joint diseases. When one compares the sequences of bacterial, mammalian and human 60kD hsp, the high homology of proteins from these different species is striking. At the same time, the 60kD hsp is an immunodominant antigen for CD4+ T cells. Thus, it was suggested that T lymphocytes originally directed against hsp of tissue-invading bacteria might cross-react with self hsp from "stressed" inflamed tissues. In human Yersinia-induced reactive arthritis, a synovial CD4+ T cell clone that had originally been shown to proliferate in response to *Y. enterocolitica*, was specific for a highly conserved hsp 60 epitope on the Yersinia and the human 60kD hsp.[11] Meanwhile, it has become obvious that at least in human rheumatoid and reactive arthritis, the pathogenetic role of the 60kD hsp as a dominant T-cell antigen in the synovial fluid was overestimated after the first studies with the recombinant protein.[12] Alternatively, recently published data on preimmunization experiments with mycobacterial hsp60 prior to induction of adjuvant arthritis with mycobacteria in complete Freund's adjuvant have led to the conclusion that the activation of T cells recognizing cross-reactive self epitopes might prevent development of chronic destructive arthritis.[13] The relevant epitope recognized by such cross-reactive epitopes must be residues 256-265 of the hsp60 molecule. An exciting extension of these animal model findings was the report that two CD4+ T cell clones from a patient with juvenile chronic arthritis recognized cross-reactive epitopes in that highly conserved region (residues 243-265), and that the donor patient indeed had a favorable outcome of the disease.[14] De Graeff Maeder et al.[15] made the interesting observation that reactivity to human hsp60 was mostly, but not exclusively, found in patients with a remitting course of juvenile chronic arthritis, i.e. the subgroup of HLA-B27-negative patients with disease of oligoarticular onset. The authors conclude from these data that self-reactive T cells sometimes mediate protection rather than autoimmune disease.

Besides $\alpha\beta$-TCR$^+$ CD4$^+$ lymphocytes, $\gamma\delta$-TCR+ lymphocytes are also known to respond to hsp60. The *in vitro* stimulation of synovial fluid mononuclear cells with either live enterobacteria like *Yersinia*, *Salmonella* and *Pseudomonas* or with purified recombinant mycobacterial or human 60kD hsp leads to a preferential expansion of γδ-TCR$^+$ lymphocytes (up to 97% of all CD3$^+$ cells after 15 days of *in vitro* culture). In contrast, heat-killed enterobacteria or other recall antigens like tetanus toxoid stimulated and expanded CD4$^+$ T cells.[16] Peripheral blood mononuclear cell stimulation with hsp also leads to a slight increase of γδ-TCR$^+$ lymphocytes in the T cell lines, but not to the extent that had been demonstrated for SF cultures (maximum 21.8% of total CD3$^+$ cells). The finding that γδ-TCR$^+$ cells from the same SF mononuclear cell population are preferentially activated by live bacteria as well as by bacterial or self hsp suggests that a common mechanism is involved in the selective expansion of these cells. One can speculate that γδ-TCR$^+$ cells play a role in the early defense mechanisms directed against cells infected by intracellular bacteria or against autologous inflamed "stressed" tissues expressing self hsp.

Cross-reacting cytotoxic lymphocytes (CTL) recognizing bacterial and self peptides

In the HLA-B27-associated spondylarthropathies, the so-called "arthritogenic peptide model"[17] concerns the possibility that molecular mimicry between bacterial and self peptides, both presented on HLA-B27, could account for autoreactivity after bacteria-induced CD8+ T cell triggering. Until today, there is still no direct evidence for the presence of such cross-reacting CTL recognizing both bacterial and self peptides in the context of HLA-B27. However, our group has recently isolated and further characterized HLA-B27-restricted anti-yersinia and anti-salmonella CTL as well as HLA-B27-restricted autoreactive CTL from the synovial fluid of patients with reactive arthritis and ankylosing spondylitis.[18] Thus, it can also be proposed that infection with *Yersinia* or *Salmonella* and the consecutive acute synovitic inflammation provides immunogenic endogenous peptides to prime self-reactive CTL without the requirement for cross-reacting peptides. Possible candidates for such endogenous peptides are heat shock protein (hsp)-derived peptides which are known to be overexpressed during target cell infection with intracellular bacteria. Hsp-derived peptides and specific CTL could maintain the inflammatory process even after the bacterial pathogen itself had been eradicated by antibacterial immune responses. HLA-B27 restricted selfreactive CTL which are likely to belong to the physiological autoreactive T cell repertoire will then show clonal expansion after temporary upregulation of former "cryptic" self determinants on synovial antigen presenting cells. The local inflammatory milieu and γ-interferon producing CD4+ T helper type 1 (Th1) cells could provide considerable amounts of γ-interferon to induce MHC-dependent changes in proteasome activities, thus favoring the production of (autoantigenic) peptides appropriate for binding to HLA-B27 and enhancing the efficiency of antigen presentation (for a review, see ref. 19). In the meantime, we have extended our previous studies and generated a larger number of HLA-B27-restricted CD8+ CTL clones with specificity for *Y. enterocolitica* or autoantigen from 3 ReA synovial fluids and PBL of 1 HLA-B27+ healthy donor. Sequencing of the TCR Vβ chains revealed that 25 TCR Vβ chains from HLA-B27-restricted clones and 2 two-clonal lines used a highly limited set of related Vβ genes with preferential rearrangement of 3 closely related Vβ families 13, 14, and 17.[20] Rearranged Vβ segments displayed a 50-85% amino acid sequence homology. The preferential usage of Vβ families 13, 14, and 17 was most obvious among synovial CTL since only 1 of 12 HLA-B27 restricted synovial CTL used Vβ6, i.e. a Vβ family other than Vβ 13, 14, or 17.

In a paper by Scofield et al.[21] it is reported that HLA-B27, but not other HLA-B proteins, share a high number of identical hexapeptide and pentapeptide sequences with proteins from gram-negative enterobacteria, suggesting convergent evolution. On this basis, presentation of enterobacterial peptides mimicking B27-specific sequences to T cells by HLA-B27 could lead to breakdown of self-tolerance to the HLA-B27 antigen and consecutive autoimmunity. In a following study,[22] the authors demonstrated that a nonapeptide from the third hypervariable region of HLA-B27 (LRRYLENGK) bearing the HLA-B27-binding motif can itself bind to HLA-B27 as well as to other sequence analogous nonapeptides derived from *Pseudomonas aeruginosa*, *Klebsiella*-nitrogenase und *Bacillus megaterium*. In the future, this hypothesis – which is a variant of the arthritogenic peptide model – needs to be tested by analyzing the immunogenic potential of these self and bacterial peptides if presented by HLA-B27, and the putative autoreactivity of the B27-restricted CD8+ CTL directed against those peptides.[23]

Superantigens

The role that bacterial superantigens play in the host-pathogen interaction remains largely undefined. However, it is clear that *in vivo* exposure to superantigens can result in significant alterations in host T cell functions and numbers. Depending on the superantigen and on the time course of exposure, this interaction can result in T cell activation, T cell depletion or anergy. One current hypothesis, which has been put forward by some investigators, suggests that one powerful consequence of exposure to superantigens is the development of autoimmunity which would result from the expansion of normally suppressed autoreactive T cell clones. In the light of this hypothesis and of the association of T cell mediated immunity in Yersinia-reactive arthritis, it is of interest that *Y. enterocolitica* produces antigenic material that has properties consistent with those of T cell superantigens. Stuart and Woodward[24] have provided data showing that in BALB/c mice T cells bearing Vβ 3, 6 and 11, and possibly 7 and 9, were expanded after exposure to *Y. enterocolitica* antigen preparations. However, in the human immune system, such superantigenic effects of *Y. enterocolitica* have not yet been demonstrated. In contrast, superantigenic effects of certain streptococcal M protein preparations (pepM5, pepM6 and pepM24) may contribute to autoimmune pathology in post-streptococcal disease. The proliferation of human T cells to pepM5 is MHC class II dependent but not MHC haplotype-restricted,[25] indicating superantigen-induced reactions. However, the superantigenicity of pepM5 has recently been reported to be due to contamination with pyrogenic exotoxins of *S. pyogenes*.

Several recent studies have suggested an involvement of both activated T lymphocytes and group A streptococci in the pathogenesis of psoriasis and also psoriatic arthritis. In acute psoriasis guttata, Leung et al.[26] showed that lesional T cells were stimulated by the streptococcal superantigen streptococcal pyrogenic exotoxin C with a consecutive *in vivo* expansion of TCR Vβ2-positive T cells.

Conclusions

In most diseases supposedly due to autoimmunity, the relevant target autoantigens have not yet been identified. In animal models as well as in various human forms of autoimmune disease, there is strong evidence that bacteria can induce or maintain autoimmune reactions. Among the mechanisms involved in the defense against bacteria and in the pathogenesis of bacteria-triggered autoimmunity, the molecular mimicry hypothesis has been widely discussed in the literature. A modification of this theory supposes that homologous bacterial and self peptides are presented on HLA class I molecules and then recognized by cross-reactive cytotoxic T lymphocytes (CTL). One critical antigen involved in the induction of cross-reactivity is the 60kD hsp which is known to be expressed by intracellular bacteria and also by stressed autologous (inflamed) tissues. Finally, under certain conditions of host and microbial virulence, bacterial endotoxins and bacterial superantigens are likely candidates to contribute to bacteria-induced autoimmune disease.

References

1 Robinson JH, Kehoe MA. Group A streptococcal M proteins: virulence factors and protective antigens. Immunol Today 1992;13:362-7.

2 Rose NR, Herskowitz A, Neumann DA, Neu N. Autoimmune myocarditis: a paradigm of post-infection autoimmune disease. Immunol Today 1988;9:117-20.

3 Shoenfeld Y, Isenberg DA. Mycobacteria and autoimmunity. Immunol Today 1988;9:178-82.

4 Schwimmbeck PL, Yu DTY, Oldstone MB. Autoantibodies to HLA-B27 in the sera of HLA-B27 patients with ankylosing spondylitis and Reiter's syndrome: Molecular mimicry with Klebsiella pneumoniae as a potential mechanism of autoimmune disease. J Exp Med 1987;166:173-81.

5 Stieglitz H, Fosmire S, Lipsky P. Identification of a 2-MD plasmid from shigella flexneri associated with reactive arthritis. Arthr Rheum 1989;32:937-46.

6 Lahesmaa R, Skurnik M, Vaara M, Leirisalo-Repo M, Nissilä M, Granfors K, Toivanen P. Molecular mimickry between HLA B27 and Yersinia, Salmonella, Shigella and Klebsiella within the same region of HLA alpha1-helix. Clin Exp Immunol 1991;86:399-404.

7 Kapasi K, Chui B, Inman RD. HLA-B27/microbial mimicry: an in vivo analysis. Immunology 1992;77:456-61.

8 Cicz RM, Urban RG, Lane WS, Gorga JC, Stern LJ, Vignali DA, Strominger JL. Predominant naturally processed peptides bound to HLA-DR1 are derived from MHC-related molecules and are heterogeneous in size. Nature 1992;358:764-7.

9 Van Rooijen N. Are bacterial endotoxins involved in autoimmunity by $CD5^+$ ($Ly\text{-}1^+$) B cells? Immunol Today 1989;10:334-6.

10 Karlsson-Parra A, Söderström K, Ferm M, Ivanyi J, Kiessling R, Klareskog L. Presence of human 65 kD heat shock protein (hsp) in inflamed joints and subcutaneous nodules of RA patients. Scand J Immunol 1990;31:283-8.

11 Hermann E, Lohse AW, Van der Zee R, Van Eden W, Mayet W-J, Probst P,

Poralla T, Meyer zum Büschenfelde K-H, Fleischer B. Synovial fluid derived Yersinia reactive T-cells responding to human 65 kD heat shock protein and heat-stressed antigen presenting cells. Eur J Immunol 1991;21:2139-43.

12 Probst P, Hermann E, Meyer zum Büschenfelde K-H, Fleischer B. Multiclonal synovial T cell response to Yersinia enterocolitica in reactive arthritis. The Yersinia 61kD heat shock protein is not the major target antigen. J Infect Dis 1993;167:385-91.

13 Anderton SM, Van der Zee R, Noordzij A, Van Eden W. Differential mycobacterial 65-kDa heat shock protein T cell epitope recognition after adjuvant arthritis-inducing or protective immunization protocols. J Immunol 1994;152:3656-64.

14 Quayle AJ, Wilson KB, Li SG, Kjeldsen-Kragh J, Oftung F, Shinnick T, Sioud M, Forre O, Capra JD, Natvig JB. Peptide recognition, T cell receptor usage and HLA restriction elements of human heat-shock protein (hsp) 60 and mycobacterial 65-kDa hsp-reactive T cell clones from rheumatoid arthritis synovial fluid. Eur J Immunol 1992;22:1315-22.

15 De Graeff Mader ER, Van Eden W, Rijkers GT, Prakken BJ, Kuis W, Voorhorst Ogink MM, Van der Zee R, Schuurmann HJ, Helders PJ, Zegers BJ. Juvenile chronic arthritis: T cell reactivity to human HSP60 in patients with a favourable course of arthritis. J Clin Invest 1995;95:934-40.

16 Hermann E, Lohse AW, Mayet W-J, Van der Zee R, Van Embden JDA, Poralla T, Meyer zum Büschenfelde K-H, Fleischer B. Stimulation of synovial fluid mononuclear cells with the human 65kD heat shock protein or with live enterobacteria leads to preferential expansion of TCR-$\gamma\delta$+ lymphocytes. Clin Exp Immunol 1992;89:427-33.

17 Benjamin R, Parham P. HLA-B27 and disease: a consequence of inadvertent antigen presentation? Rheum Dis Clin N Amer 1992;18 11-21.

18 Hermann E, Yu DTY, Meyer zum Büschenfelde K-H, Fleischer B. HLA-B27-restricted CD8 T cells derived from synovial fluids of patients with reactive arthritis and ankylosing spondylitis. Lancet 1993;342:646-50.

19 Hermann E, Fleischer B, Meyer zum Büschenfelde K-H. Bacteria-specific cytotoxic CD8+ T cells: a missing link in the pathogenesis of the HLA-B27-associated spondylarthropathies. Ann Med 1994;26:365-9.

20 Duchmann R, May E, Ackermann B, Goergen B, Meyer zum Büschenfelde K-H, Märker-Hermann E. HLA-B27 restricted cytotoxic T lymphocyte responses to arthritogenic enterobacteria or self antigens are dominated by closely related TCRBV gene segments. Scand J Immunol 1996;43:101-8.

21 Scofield RH, Warren WL, Koelsch H, Harley JB. A hypothesis for the HLA-B27 immune dysregulation in spondylarthropathy: Contributions from enteric organsims, B27 structure, peptides bound by B27, and convergent evolution. Proc Natl Acad Sci USA 1993;90:9330-4.

22 Scofield RH, Kurien B, Gross T, Warren WL, Harley JB. HLA-B27 binding of peptide from its own sequence and similar peptides from bacteria: implications for spondylarthropathies. Lancet 1995;345:1542-4.

23 López de Castro J. Structure, function, and disease association of HLA-B27. Curr Opin Rheumatol 1994;6:371-7.

24 Stuart PM, Woodward JG. Yersinia enterocolitica produces superantigenic

activity. J Immunol 1992;148:225-33.
25 Tomai MA, Korb M, Majumdar G, Beachey EH. Superantigenicity of streptococcal M protein. J Exp Med 1990;174:359-62.
26 Leung DYM, Travers JB, Giorno R, Norris DA, Skinner R, Aelion J, Kazemi LV, Kim MH, Trumble AE, Kotb M, Schlievert PM. Evidence for a streptococcal superantigen-driven process in acute guttate psoriasis. J Clin Invest 1995;96:2106-12.

ULRICH FEIGE AND JILL GASSER

Heat shock proteins, autoimmune disease and antigenic mimicry

Introduction

Susceptiblity to autoimmune disease(s) is determined by genetic factors and several genes are involved.[1] Environmental factors also play a role in induction of autoimmune disease (see, for example, refs. 2 & 3). Diseases in animals are used as models for (similar) diseases in human beings. Autoimmune disease can be induced in animals by an injection of (auto-)antigen in adjuvant, although there are a few 'spontaneous' autoimmune diseases in animals such as diabetes in NOD mice or BB rats.[4] The autoantigens for most autoimmune diseases are not identified yet; however, for any given autoimmune disease, an immune response against several autoantigens may result in the same disease pattern (see, for example, refs. 5 & 6). In principle, autoimmune disease has only two prerequisites: (i) an immune response against an antigen, and (ii) the presence of the respective antigen in an organ. For example, an episode of arthritis in a joint is induced by simultaneous injection of an antigen intravenously and T cells specific for this antigen intra-articularly, or vice versa (U. Feige et al., unpublished). Other examples have been demonstrated and were called bystander arthritis.[7] Another example worth mentioning in this context is the so-called antigen-induced arthritis. An immune response to bovine serum albumin (BSA) is induced by subcutaneous immunization of mice, rats or guinea pigs with BSA in adjuvant. When, some weeks later, the animals are challenged intra-articularly with BSA, an arthritis develops which has features of an autoimmune arthritis. However, this arthritis is self-limiting, presumably due to the fact that the driving 'autoantigen' BSA is not present in the joint for a long time. Upon intra-articular rechallenge with BSA a new episode of arthritis is induced.[8] Again, this results in self-limiting disease. In other words, for an autoimmune disease to be started or to flare up requires only the presence and accessibility of an (auto-)antigen and an immune response directed to this (auto-)antigen.

Antigenic mimicry is understood to be the presence of a (similar) epitope on two different antigens.[9] Antigenic mimicry may be a basis for the influence of opportunistic infections on autoimmunity and autoimmune disease.[3,10,11] However, the influence of infections on autoimmunity (to hsp) may not neccessarily result in autoimmune disease (see ref. 12).

In this article, potential mechanisms of immune regulation in the induction and (self-)limitation of autoimmune disease will be discussed.

A role for opportunistic infection(s) in induction of autoimmune disease

Infections with *Streptococci, Chlamydia, Klebsiella, Borrelia* and other bacteria can result in an arthritic episode or even in chronic arthritis. It is now generally accepted that in reactive arthritis, the immune response of the host during infection is reactive with an antigen in the joints, resulting in precipitation of disease in the joint(s).[13,14] It has been previously speculated that persistence of bacteria or bacterial antigens in joints is responsible for the perpetuation of the disease process.[11,15] However, it is also posssible that the immune reaction induced to fight infection is cross-reactive with an autoantigen in the joint (see, for example, refs. 16 & 17), but in no case has the autoantigen been identified so far. Based on the above, the concept was developed that immunity to bacterial antigens develops after birth during opportunistic infections and that by cross-reactivity with autoantigens this immunity turns into autoimmunity which results in autoimmune disease (see below).

Autoimmunity to heat shock proteins is ubiqitous

In the late eighties, two observations were made. One was that the 'common antigen' of bacteria is a member of the bacterial hsp60 family[18,19] and the other was that the member of the hsp60 family in Mycobacteria, hsp65, is involved in adjuvant arthritis (AdA) in Lewis rats.[20] AdA is induced by an injection of a suspension of mycobacteria in oil intradermally at the base of the tail or subplantarily. T cell clones A2b and A2c respond to mycobacterial hsp65 in vitro and induce or prevent arthritis in Lewis rats, respectively (see ref. 20).

Besides the α,ß-T cells a second set of T cells, γ,δ-T cells, exists. A high proportion of γ,δ-T cells respond to mycobacterial hsp65 as their antigen in vitro.[21,22] Immune responses of both α,ß- and γ,δ-T cells to mycobacterial hsp65 and autologous hsp60 were found in cord blood and neonatally.[23-25] Interestingly, in 6-week-old Lewis rats which were housed under germ-free conditions, T cell responses to mycobacterial antigens were comparable to those of littermates housed under conventional conditions (Gasser and Feige, unpublished observations). One conclusion which can be drawn from the above is that autoimmunity to hsp is present at and around birth. Part of this autoimmunity might be cross-reactive with bacterial antigens such as hsp60 (common antigen or GroEL). Immune responses to bacterial antigens might stimulate the natural immunity to hsp and other autoantigens. This is in contrast to the previously held view that immunity to bacterial antigens might cross-react with autoantigens and ultimately result in autoimmunity (Figure 1).

Autoimmune disease can be modulated with mycobacterial hsp65

When hsp65 was found as a T cell antigen involved in AdA, attempts were made to induce AdA with mycobacterial hsp65. However, it was found that hsp65 does not induce AdA but rather a state of resistance to AdA.[20,32,33] Thereafter, in studies using overlapping peptides, the epitope for T cell clones A2b and A2c was found to be contained in the amino acid sequence 180 to 188 of mycobacterial hsp65 (peptide 180-188).[34] As had been shown with whole mycobacteria and purified mycobacterial hsp65, preimmunization with peptide 180-188 resulted in an AdA resistant state in Lewis rats.[27,33,35,36] Interestingly, this AdA resistance was accompanied by a higher T cell immune response to mycobacterial antigens.[33,37] An analysis of the T cell response in mycobacterially immunized versus arthritic rats revealed different epitope specificity patterns.[37] Immunization of rats with these epitopes before an arthritogenic challenge with mycobacteria showed that peptide 256-265 of the mycobacterial hsp65 is a second epitope which can induce resistance to AdA, indi-

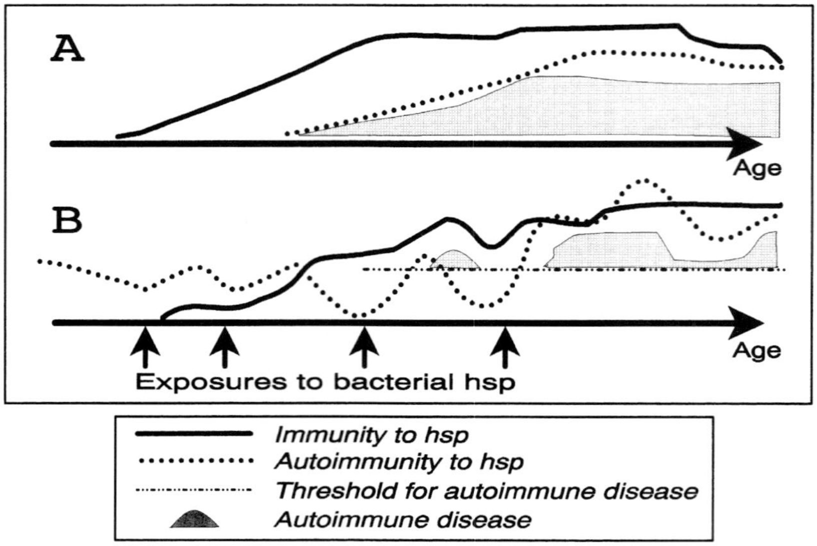

FIGURE 1 Immunity to bacterial hsp60 and autoimmunity to self hsp60. Hsp are highly conserved molecules which fulfil functions essential for cell survival. It was therefore surprising when 'common antigen', the most immunodominant protein of bacteria, was found to be a member of the bacterial hsp60 family.[18,19] In addition, autoimmunity to hsp60 was detected at birth and in normal individuals.[23,24] It is therefore no longer unexpected that immune responses to hsp60 are also involved in autoimmune diseases. However, it is important to note that autoimmune diseases can be modulated even therapeutically by vaccinations with mycobacterial hsp65 or self hsp60.[26,27] Immunity, autoimmunity and autoimmune disease may be directed to identical epitopes. However, there is also autoimmunity without autoimmune disease.[12,28-31]
A. The 'old' view: Immunity to bacterial hsp develops during bacterial infections and at some point in life, due to cross-reactivity, turns into autoimmunity. Autoimmune disease is the 'logical' and inevitable consequence of this autoimmunity and occurs when hsp60 is overexpressed in a tissue or organ or when a cross-reactive epitope is present on a molecule different from hsp in a tissue or organ (molecular mimicry).
B. The 'new' view: Autoimmunity to hsp 60 is present at birth. This autoimmunity is modulated during bacterial infections and may lead to autoimmune disease. However, autoimmune disease is not an inevitable sequela of autoimmunity. Regulatory processes within the network of T cell reactions may downregulate autoimmune disease. In fact, intervention in autoimmune disease using vaccinations with peptides derived from hsp60 resulted in a downregulation of the inflammatory component of the autoimmune disease (see text).

cating that several 'disease-protecting' epitopes can be present in one protein.

Early on, it had been recognized that immunization with mycobacterial hsp65 can prevent autoimmune disease in several autoimmune disease models in animals, among which are models of arthritis in rats and mice (such as AdA, streptococcal cell wall induced arthritis, pristane induced arthritis) but also diabetes (in NOD mice) [for a review, see ref. 38]. NOD mice develop a spontaneous insulitis which results in overt diabetes at the age of about 4 months.[39] In NOD mice, immunization with hsp60 in incomplete Freund's adjuvant leads to an episode of insulitis and diabetes from which the mice recover and after which they do not develop the spontaneously occurring diabetes. Interestingly, a 'less immunogenic' immunization with hsp60 in saline results in protection against the spontaneously occurring diabetes without inducing an episode of insulitis and diabetes.

It is noteworthy that in a clinical trial with newly diagnosed insulin dependent diabetes mellitus patients, 17 of 27 went into remission after treatment with BCG (*Mycobacterium tuberculosis*).[40] However, treatment with whole BCG is not without risk of inducing autoimmune disease. Arthritis has been observed as a side effect of BCG therapy in bladder cancer patients.[41]

From these results it appeared that immunity to hsp65 (and hsp60) was involved in several autoimmune diseases where it was difficult to believe that hsp65 had any role as a primary antigen. Rather, it could be concluded that there is a network within the T cell compartment which has hsp65 (hsp60) as one of the target antigens. Following from this, it was postulated that a panel of antigens serving T cells in a network was involved in autoimmunity and autoimmune disease. Autologous hsp60 may be one of the more common players in the regulatory T cell network in autoimmunity and autoimmune disease.[16,28-31,38,42,43]

Autoimmune disease can be modulated with several (auto-)antigens

The question of why autoimmune diseases such as arthritis or diabetes occur may be answered by the presence of a cross-reactive antigen in joints or the pancreas, etc. In other words, the autoantigen targets the (auto-)immunity to that organ and precipitates autoimmune disease. However, what are these autoantigen(s)? In collaboration with Jürgen Mollenhauer we identified a chondrocyte protein with an approximate molecular weight of 65 kD (CH65) as a potential autoantigen in AdA. The kinetics of the cellular immune responses to CH65 and mycobacterial antigens parallel each other during the course of AdA (unpublished observations). In addition, it has been shown that pretreatment of rats with CH65 retards the onset of AdA similar to pretreatment with mycobacterial hsp65.[17]

In diabetes in NOD mice an array of antigens have been implicated.[44] One of these antigens, GAD65, is believed to be the first antigen involved.[44-46] Treatment of NOD mice with this autoantigen resulted in suppression of the diabetes,[45] again comparable to pretreatment with mycobacterial hsp65.[39]

The following conclusions can be drawn from the results presented so far:

(i) several antigens/autoantigens may be involved in the underlying immune reactions in an autoimmune disease,
(ii) autoantigen(s) can be used to modulate disease,
(iii) autoimmune disease can be modulated with a foreign antigen involved in the immunological network of responses, such as hsp, and therefore
(iv) the autoantigen need not necessarily be known to attempt treatment of autoimmune disease by antigen-induced modulation of the underlying autoimmune response.

Peptide based immunotherapy of autoimmune disease

It has been summarized above that disease (AdA in rats, diabetes in NOD mice) can be prevented with mycobacterial hsp65 or mouse hsp60 or peptides thereof, or with autoantigens such as CH65 or GAD65.[17,45] However, as mentioned at the beginning of this article, not all individuals who are prone to autoimmune disease necessarily will develop disease. Therefore, vaccination programs comparable to those used for prevention of infectious diseases may not be feasible or advised for autoimmune diseases. This situation led to the question: Can the above principles of hsp-based vaccination be applied in a therapeutic regimen? The answer is yes. A few examples in animal models have been described:

Rats with AdA, when treated after onset of disease with 0.1 or 1 mg of peptide 180-188 of mycobacterial hsp65, show no or reduced further paw swelling and weight loss.[27] Radiographs were taken from these rats 35 days after induction of disease with mycobacteria.

Massive destruction of joints and bone in paws was apparent in the AdA control group whereas peptide 180–188 treated rats showed no or much less destruction.[27]

NOD mice develop insulitis and hyperglycemia. It is interesting to note that NOD mice can be protected from developing diabetes by immunization with peptide p277, which corresponds to sequence 236 to 260 of the mouse hsp60.[26,39,47] The sequence 236 to 260 is highly homologous in the hsp60/hsp65 of mice and mycobacteria. In the experiment reported by Elias and Cohen[26] all control NOD mice died before they reached 42 weeks of age. Impressively, NOD mice can be rescued from their diabetes-induced death by a single immunization with peptide p277, even when administered as late as at 17 weeks of age.[26]

These results are interesting because they show that even in severe autoimmune disease the immune response is the driving force, also of the inflammatory component. Furthermore, the immune response can be modulated by (peptides derived from) autoantigen or hsp60, indicating that autoimmunity contains a component for downregulation of autoimmune disease.[12,16,28-31,38,42] As a consequence, mere immunosuppressive treatment of autoimmune disease has to be questioned because it may also affect the arm of autoimmunity which is needed for protection from autoimmune disease (see below).

Immunosuppression versus immunomodulation in the treatment of autoimmune diseases

Immunosuppressive treatment with steroids, azathiaprine, methotrexate, cyclophosphamide or cyclosporin A is commonly used in autoimmune diseases. However, this type of treatment may be reconsidered in view of the finding that autoimmune diseases might be treatable by immunomodulation. For example, clinical trials are in process in rheumatoid arthritis with an extract of *E.coli*, Subreum[R], which contains hsp60, GroEL, as a main constituent (for a review, see ref. 48). In addition, 'paradoxical effects' of immunosuppressive treatment have been demonstrated. Kaibara and colleagues[49] have used cyclosporin A (CsA) in type II collagen-induced arthritis in rats and demonstrated that the time of treatment with CsA is critical for the outcome. Treatment with CsA from day 0 - 7 or 0 - 13 after immunization with collagen resulted in a reduction of disease whereas treatment from day 7 - 13 or 14 - 20 resulted in an exacerbation of disease. Similar results were obtained in type II collagen-induced arthritis in DBA/1 mice by treatment with anti-α,ß-T cell receptor antibody.[50] It is important to note that when immunosuppressive treatment is or must be stopped, a flare of the autoimmune disease usually occurs. Therefore, it might be advisable to emphasize exploration of the possibilities of immunomodulation for the benefit of the patients suffering from autoimmune disease. Especially since immunotherapy of autoimmune diseases based on (therapeutic) vaccination with peptides appears to be a much safer and gentler treatment and results in a disease-resistant state maintained by the immune system itself.[12,42]

Molecular mimicry

Molecular mimicry is defined as the existence of an epitope on both a foreign and an autoantigen.[9] Molecular mimicry has been studied in both viral and autoimmune diseases (see reactive arthritis above). In view of the regulatory networks in T cell (auto-)immunity which have been discussed above, molecular mimicry appears to result not only in 'dangerous' immune reactions[51] but may also provide a basis for immune homeostasis.[12,16,28-31,38] Analyses of homologies between proteins are quite helpful in the search for identical or similar epitopes. Computer searches through sequence databases demonstrate that identical epitopes on different antigens are not uncommon (see ref. 52). Of course, not every homology of epitopes detected in databases has impact on the immune system. Only those epitopes which are produced and stay intact during antigen processing and presentation are potential candidates. Antigenic mimicry can occur between two foreign antigens, between a foreign antigen and an autoantigen, and also between two autoantigens. Therefore, it appears natural that the immune system has evolved to deal with such situations and has acquired the ability to avoid responses which lead to autoimmune disease. A network of T cells directed against a few (auto-)antigens seems to be a good choice, because it gives the immune system the capacity to downregulate T cell responses which may result in autoimmune disease. Induction of oral tolerance for an antigen[48,53] may also enhance the suppressive arm of an immune response. The latter appears to be of great importance given the variety of antigens the gut is exposed to.

On the contrary, induction of autoimmune disease 'forced' by injection of peptides or antigens in a strong adjuvant may simply overcome the natural protective mechanisms. It is interesting to note that many epitopes can be used to induce autoimmune disease. For example, a number of 'encephalitogenic' epitopes of myelin basic protein have been described for the Lewis rat (see ref. 5). Encephalomyelitis can also be induced by myelin proteolipid protein,[6] i.e., several (auto-)antigens can induce an identical disease pattern. Indications for the existence of downregulatory (T suppressor?) cells also come from the fact that immunosuppressive treatment given to mice with type II collagen-induced arthritis 'at the wrong time' results in exacerbation of disease.[49,50]

During the last few years, we have begun to understand some aspects of the autoregulatory network of T cells in autoimmunity. The view that antigenic mimicry between a foreign antigen and an autoantigen leads directly to autoimmune disease has had to be revised (Figure 1). Rather, we perceive antigenic mimicry as the basis for immune homeostasis and the prevention of autoimmune disease(s). Knowledge of the involvement of a few antigens such as the mycobacterial hsp65 in autoimmunity and autoimmune disease allows us to interfere with the course of an autoimmune disease.[38] Importantly, the induction of resistance to a given autoimmune disease is not necessarily correlated to induction of tolerance in the immunological sense to an (auto-)antigen (see refs. 12, 16, 29, 38, 42, 43). It appears that nature has developed a balanced way to respond to antigens and autoantigens to prevent autoimmune disease. Even overt autoimmune disease may be susceptible to the action of counter-regulatory loops which revert autoimmune disease 'back' to autoimmunity. This may be an explanation for spontaneous remissions of autoimmune diseases.[12,42]

Outlook

The possibilities for the treatment of autoimmune diseases based on knowledge of target antigens in the T cell network of autoimmunity are intriguing. Especially when treatment is based upon epitopes which are not involved in the autoimmune disease-inducing process, peptide-based 'therapeutic vaccinations' may be safe and free of side effects. At least from the results obtained so far in animal models of autoimmune disease, it appears that there may be more epitopes which lead to resistance against autoimmune disease than epitopes inducing autoimmune disease. In addition, immunization schedules may be chosen which are known rather to induce resistance against autoimmune disease. It appears that antigens such as hsp65/hsp60 may be used for the treatment of several autoimmune diseases, even though the inciting (auto-)antigen for a given disease is not known.

Summary

Immune responses to an autoantigen do not necessarily result in autoimmune disease. We have learned to distinguish autoimmunity from autoimmune disease. Research during recent years has led to an understanding of T cell autoimmunity as a network driven by a panel of (auto-)antigens. Within this network, at least two arms of T cell immunity exist: an autoimmune disease inducing and an autoimmune disease preventing arm. Autoimmune disease occurs when the autoimmmune disease inducing arm is strengthened. This is achieved by immunization of animals with antigen in a strong adjuvant such as complete Freund's adjuvant. Conversely, autoimmune disease can be prevented or treated by using the identical antigens and applying them in such a way that the autoimmune disease preventing arm is activated.

One of the molecules which is part of the network of antigens in autoimmunity and autoimmune disease is (mycobacterial) heat shock protein 65 (hsp65). Immunization of animals with hsp65 prevents disease in a number of animal models. A whole array of so-called 'induced' arthritic diseases in mice or rats, as well as the spontaneously occurring diabetes in NOD mice, can be prevented or treated with hsp65. It is noteworthy that it is no longer necessary to know the first antigen involved in the process leading to an autoimmune disease because the disease process can apparently be influenced by any of a number of antigens participating in the network.

References

1 Carson DA. Genetic factors in the etiology and pathogenesis of autoimmunity. FASEB J 1992;6:2800-5.

2 Bernard CC, Kerlero de Rosbo N. Multiple sclerosis: an autoimmune disease of multifactorial etiology. Curr Opin Immunol 1992;4:760-5.

3 Toivanen P, Toivanen A, Brines R. When is an autoimmune disease not an

autoimmune disease? Immunol Today 1994;15:556-9.
4 Birk OS, Cohen IR. T-cell autoimmunity in type 1 diabetes mellitus. Curr Opin Immunol 1993;5:903-9.
5 Vandenbark AA, Vainiene M, Celnik B, Hashim GA, Buenafe A, Offner H. Definition of encephalitogenic and immunodominant epitopes of guinea pig myelin basic protein (Gp-BP) in Lewis rats tolerized neonatally with Gp-BP or Gp-BP peptides. J Immunol 1994;153:852-61.
6 Tuohy VK. Peptide determinants of myelin proteolipid protein (PLP) in autoimmune demyelinating disease: a review. Neurochem Res 1994;19:935-44.
7 Holoshitz J, Naparstek Y, Ben-Nun A, Marquardt P, Cohen IR. T lymphocyte lines induce autoimmune encephalomyelitis, delayed hypersensitivity and bystander encephalitis or arthritis. Eur J Immunol 1984;14:729-34.
8 Buchner E, Braeuer R, Schmidt C, Emmrich F, Kinne RW. Induction of flare-up reactions in rat antigen-induced arthritis. J Autoimmun 1995;8:61-74.
9 Oldstone MB. Molecular mimicry as a mechanism for the cause and a probe uncovering etiologic agent(s) of autoimmune disease. Curr Top Microbiol Immunol 1989;145:127-35.
10 Cohen IR, Young DB. Autoimmunity, microbial immunity and the immunological homunculus. Immunol Today 1991;12:105-10.
11 Toivanen P, Toivanen A. Role of micro-organisms in the pathogenesis of arthritis: lessons from reactive and Lyme arthritis. Scand J Rheumatol Suppl 1995;101:191-7.
12 Feige U, Van Eden W. Infection, autoimmunity and autoimmune disease. In: Feige U, Morimoto R, Yahara I, Polla BS. Stress-inducible cellular responses. Basel: Birkhäuser, 1996:359-73.
13 Hermann E. T cells in reactive arthritis. APMIS 1993;101:177-86.
14 Probst P, Hermann E, Fleischer B. Role of bacteria-specific T cells in the immunopathogenesis of reactive arthritis. Trends Microbiol 1994;2:329-32.
15 Merilahti-Palo R, Pelliniemi LJ, Granfors K, Söderström KO, Von Essen R, Similae A, Toivanen A. Electron microscopy and immunolabeling of Yersinia antigens in human synovial fluid cells. Clin Exp Rheumatol 1994;12:255-9.
16 Feige U. Heat-shock proteins and arthritis. In: Immunology of the joint and the eye. Zierhut M, Saal J, Thiel HJ, editors. Buren: Aeolus Press, 1995:47-61.
17 Feige U, Schulmeister A, Mollenhauer J, Brune K, Bang H. A constitutive 65 kDa chondrocyte protein as a target antigen in adjuvant arthritis in Lewis rats. Autoimmunity 1994;17:233-9.
18 Shinnick TM, Vodkin MH, Williams JL. The *Mycobacterium tuberculosis* 65 kDa protein is a heat shock protein which corresponds to common antigen and to the *E.coli* groEL protein. Infect Immun 1988;56:446-51.
19 Thole JER, Hindersson P, De Bruyn J, Cremers F, Van der Zee R, De Cock H, Tommassen J, Van Eden W, Van Embden JD. Antigenic relatedness of a strongly immunogenic 65 kDa mycobacterial protein antigen with a similarly sized ubiquitous bacterial common antigen. Microbiol Pathog 1988;4:71-83.
20 Van Eden W, Thole JE, Van der Zee R, Noordzij A, Van Embden JD, Hensen EJ, Cohen IR. Cloning of the mycobacterial epitope recognized by T lymphocytes in adjuvant arthritis. Nature 1988;331:171-3.
21 Born W, Hall L, Dallas A, Boymel J, Shinnick T, Young D, Brennan P,

O'Brien R. Recognition of a peptide antigen by heat shock-reactive γ,δ T lymphocytes. Science 1990;249:67-9.

22 Born W, Happ MP, Dallas A, Reardon C, Kubo R, Shinnick T, Brennan P, Born W. Recognition of heat-shock proteins and γ,δ cell function. Immunol Today 1990;11:40-3.

23 Iwasaki A, Yoshikai Y, Yuuki H, Takimoto H, Nomoto K. Self-reactive T cells are activated by the 65-kDa mycobacterial heat-shock protein in neonatally thymectomized mice. Eur J Immunol 1991;21:597-603.

24 O'Brien RL, Fu YX, Cranfill R, Dallas A, Ellis C, Reardon C, Lang J, Carding SR, Kubo R, Born W. Heat shock protein hsp60-reactive gamma delta cells: a large, diversified T-lymphocyte subset with highly focused specificity. Proc Natl Acad Sci USA 1992;89:4348-52.

25 Fischer HP, Sharrock CE, Panayi GS. High frequency of cord blood lymphocytes against mycobacterial 65-kDa heat-shock protein. Eur J Immunol 1992;22:1667-9.

26 Elias D, Cohen IR. Peptide therapy for diabetes in NOD mice. Lancet 1994;343:704-6.

27 Feige U, Gasser J. Therapeutic intervention with mycobacterial heat shock protein peptide 180-188 in adjuvant arthritis in Lewis rats. Mediation Inflammation 1994;3:304.

28 Cohen IR. Natural id-anti-id networks and the immunological homunculus. In: Atlan H, Cohen IR, editors. Springer Series in Synergetics, Vol. 46. Berlin/Heidelberg: Springer-Verlag, 1989:6-12.

29 Cohen IR. Autoimmunity to chaperonins in the pathogenesis of arthritis. Annu Rev Immunol 1991;9:567-89.

30 Cohen IR. Autoimmunity to hsp65 and the immunologic paradigm. Adv Intern Med 1992;37:295-311.

31 Cohen IR. Autoimmunity shifts paradigms. Isr J Med Sci 1994;30:37-8.

32 Billingham ME, Carney S, Butler R, Colston MJ. A mycobacterial 65-kD heat shock protein induces antigen-specific suppression of adjuvant arthritis, but is not itself arthritogenic. J Exp Med 1990;171:339-44.

33 Yang XD, Gasser J, Feige U. Prevention of adjuvant arthritis in rats by a nonapeptide from the 65-kD mycobacterial heat shock protein: specificity and mechanism. Clin Exp Immunol 1992;87:99-104.

34 Van der Zee R, Van Eden W, Meloen RH, Noordzij A, Van Embden JD. Efficient mapping and characterization of a T cell epitope by the simultaneous synthesis of multiple peptides. Eur J Immunol 1989;19:43-7.

35 Yang XD, Gasser J, Riniker B, Feige U. Treatment of adjuvant arthritis in rats: vaccination potential of a synthetic nonapeptide from the 65 kDa heat shock protein of mycobacteria. J Autoimmun 1990;3:11-23.

36 Yang XD, Gasser J, Feige U. Prevention of adjuvant arthritis in rats by a nonapeptide from the 65-kD mycobacterial heat shock protein. Clin Exp Immunol 1990;81:189-94.

37 Anderton SM, Van der Zee R, Noordzij A, Van Eden W. Differential mycobacterial 65-kDa heat shock protein T cell epitope recognition after adjuvant arthritis inducing or protective immunization protocols. J Immunol 1994;152:3656-64.

38 Feige U, Cohen IR. The 65-kDa heat-shock protein in the pathogenesis, prevention and therapy of autoimmune arthritis and diabetes mellitus in rats and mice. Springer Semin Immunopathol 1991;13:99-113.
39 Elias D, Markovits D, Reshef T, Van der Zee R, Cohen IR. Induction and therapy of autoimmune diabetes in the non-obese diabetic (NOD/Lt) mouse by a 65-kDa heat shock protein. Proc Natl Acad Sci USA 1990;87:1576-80.
40 Shehadeh N, Calcinaro F, Bradley BJ, Bruchlim I, Vardi P, Lafferty KJ. Effect of adjuvant therapy on development of diabetes in mouse and man. Lancet 1994;343:706-7.
41 Goupille P, Poet JL, Jattiot F, Mattei JP, Vedere V, Tonolli-Seabian I, Roux H, Valat JP. Three cases of arthritis after BCG therapy for bladder cancer. Clin Exp Rheumatol 1994:12:195-7.
42 Yang XD, Feige U. The 65 kDa heat shock protein: a key molecule mediating the development of autoimmune arthritis? Autoimmunity 1991;9:83-8.
43 Yang XD, Feige U. Heat shock proteins in autoimmune disease. From causative antigen to specific therapy? Experientia 1992;48:650-6.
44 Gelber C, Paborsky L, Singer S, McAteer D, Tisch R, Jolicoeur C, Buelow R, McDevitt H, Fathman CG. Isolation of nonobese diabetic mouse T cells that recognize novel autoantigens involved in the early events of diabetes. Diabetes 1994;43:33-9.
45 Tisch R, Yang XD, Liblau RS, McDevitt HO. Administering glutamic acid decarboxylase to NOD mice prevents diabetes. J Autoimmun 1994;7:845-50.
46 Tisch R, Yang XD, Singer SM, Liblau RS, Fugger L, McDevitt HO. Immune response to glutamic acid decarboxylase correlates with insulitis in non-obese diabetic mice. Nature 1993;366:72-5.
47 Elias D, Reshef T, Birk OS, Van der Zee R, Walker MD, Cohen IR. Vaccination against autoimmune mouse diabetes with a T-cell epitope of the human 65-kDa heat shock protein. Proc Natl Acad Sci USA 1991;88:3088-91.
48 Vischer TL, Van Eden W. Oral desensitation in rheumatoid arthritis (RA). Ann Rheum Dis 1994;53:708-10.
49 Kaibara N, Hotokebuchi T, Takagishi K, Katsuki I. Paradoxical effects of cyclosporin A on collagen arthritis in rats. J Exp Med 1983,158:2007-15.
50 Maeda T, Saikawa I, Hotokebuchi T, Sugioka Y, Eto M, Murakami Y, Nomoto K. Exacerbation of established collagen-induced arthritis in mice treated with an anti-T cell receptor antibody. Arthr Rheum 1994;37:406-13.
51 Matzinger P. Tolerance, danger, and the extended family. Annu Rev Immunol 1994;12:991-1045.
52 Jones DB, Coulson AFW, Duff GW. Sequence homologies between hsp60 and autoantigens. Immunol Today 1994;14:115-8.
53 Hafler DA, Weiner HL. Antigen-specific immunosuppression: oral tolerance for the treatment of autoimmune disease. Chem Immunol 1995:60:126-49.

JOHN V.FORRESTER, JANET LIVERSIDGE, ANDREW DICK,
MARIA KUPPNER AND LYNNE LUMSDEN

Experimental uveitis and autoimmunity

Introduction

Intraocular inflammation (uveitis) is classified as exogenous when a causative agent can be identified and endogenous when there is no identifiable cause and may take many clinical forms. Uveitis is also categorised according to its anatomical location[1] and pathogenetically, can be considered in two categories: anterior uveitis involving the iris and anterior ciliary body, and posterior uveitis involving the choroid/retina and the pars plana of the ciliary body. This division has considerable similarity to the broadly separate categories of arthritis, rheumatoid arthritis and reactive arthritis, in which the former is considered to be a MHC Class II-mediated delayed-type hypersensitivity (DTH) response while the latter appears to be a T cell cytotoxicity-mediated disease involving predominantly MHC Class I mechanisms.[2]

Studies of human intraocular inflammation have been restricted by the lack of adequate histological material; most tissue samples studied are of end-stage disease and are therefore not particularly informative. In addition, immunohistological studies are further hampered by the difficulties in obtaining good morphology using conventional processing techniques such as cryosectioning. Some useful information has emerged from the limited studies that are available, mostly in posterior uveitis syndromes. Conditions such as sarcoidosis and sympathetic ophthalmia appear to adopt a DTH pattern with a perivascular infiltration of CD4, and less so of CD8, T lymphocytes but a predominance of macrophages in the microgranulomatous lesions.[3,4] In other rarer entities such as the subretinal, posterior fibrosis/uveitis syndrome, B cells appear to feature in large numbers.[5] Similar studies in human anterior uveitis have not been possible.

Accordingly, many attempts have been made to develop representative, experimental models of intraocular inflammation. Broadly, these can also be categorised as exogenous and endogenous. Exogenous models involve the intraocular injection of a uveitogen, usually a foreign protein or cytokine, while endogenous experimental models involve the inoculation of putative autoantigens at a site distant from the eye.

Models of experimental uveitis

ANTERIOR UVEITIS Experimental models of exogenous uveitis have been known for many years. On the assumption that many types of anterior uveitis were somehow linked to bacterial infection, yet did not produce an endophthalmitis, it was thought that endotoxin release from bacteria accounted for the uveitic response. Early studies confirmed that intravitreal injections of bacterial endotoxin produced a marked anterior uveitis[6] and this became a standard model for studies on disease pathogenesis.[7] However, in this model, it became clear that there was a systemic effect of endotoxin since the uninoculated eye also developed uveitis, although to a lesser degree.[7] Later, it was shown that intradermal injections of endotoxin induced a characteristic anterior uveitis and the distinction between endogenous and exogenous experimental uveitis became blurred.[8]

Endotoxin-induced uveitis (EIU) is currently used extensively as a model of acute anterior uveitis. EIU is characterised by a rapid onset of inflammation with early influx of polymorphonuclear leukocytes followed by monocytes, the entire response being completed in 72-96 hrs with few ocular sequelae. Endotoxin causes release of tumour necrosis factor-α (TNFα), and interleukins-1, 6 and 8 (IL-1, 6 and 8), both systemically and into the intra-ocular compartments[9]. Some of this activity is generated by intraocular synthesis of cytokine[10] while the direct intraocular inoculation of TNFα, IL-1 and IL-6 will also induce severe uveitis.[11,12] Despite the obvious effects of TNFα, systemic administration of monoclonal antibodies to TNFα fails to inhibit the disease, indicating that the effects on local cytokine production are of paramount importance in EIU.[13]

Similarities between EIU and human acute anterior uveitis include its self-limiting nature and its general restriction to the anterior segment, despite some inflammatory cellular infiltrate in the retina. However, many human forms of anterior uveitis are associated with connective tissue disease, especially the arthritides. Models of experimental autoimmune arthritis using Type II collagen[14] have been described, suggesting a link between uveitis and arthritis since Type II collagen is a major component of cartilage and vitreous. However, a mild, low grade chronic anterior uveitis has also been described in adjuvant arthritis, and it has been suggested that the common (auto)antigen in this disorder may be cartilage proteoglycan, since homology between mycobacterial and cartilage/vitreous proteoglycan has been described (for a review, see ref. 15). Interestingly, patients with inflammatory arthritis have positive cellular immune responses to cartilage proteoglycans.[16]

Intravitreal injection of pro-inflammatory mediators such as endotoxin and cytokines is not essential to induce an exogenous uveitis. 'Inert' foreign proteins such as albumin induce a well characterised immunogenic uveitis, developing 10-14 days after injection into the vitreous. This has been shown to be induced by induction of antigen-specific B cells which localise to the eye and interact with persistent antigen trapped within the vitreous. This model is truly an immunogenic response, in this case to a foreign antigen present within the eye.[17-19] Furthermore, a marked recall response can be induced by the intravenous injection of antigen several weeks after the original uveitis has subsided. In addition to recrudescence of the uveitis, there is a marked increase in the number of local antibody producing cells in the eye.[20]

However, most forms of endogenous uveitis are considered to arise in response to autoantigens present within the eye and to which normal tolerogenic signal responses have failed or have not occurred during development, rather than to foreign antigens which have somehow gained access to the eye. The paradigm for autoimmune uveitis is sympathetic ophthalmitis (see below). Classically, human lens-induced uveitis is also considered to belong to this category of 'autoimmune' uveitic response. However, most forms of experimental (and clinical) lens-induced uveitis require, in addition to peripheral sensitisation to lens protein, a direct injury to the lens itself.[21,22] Most recently, the requirement for peripheral sensitisation combined with lens injury was further emphasised in a transgenic mouse model in which the protein, lysozyme, was expressed in the lens via the crystallin promoter. Uveitis failed to occur spontaneously in this model despite the full expression of the 'foreign' protein in the lens unless the animal was transfused with syngeneic T cells sensitised to lysozyme. It would thus appear that for the induction of endogenous autoimmune uveitis, as for other forms of organ-specific autoimmune disease,[23] peripheral sensitisation and tissue access or homing (in the case of the eye breakdown of the blood ocular barrier) are required.

Although most experimental models of acute anterior uveitis involve intraocular manipulation of some sort, recent studies have shown that a predominantly anterior uveitis can be induced by distant (intradermal) injection of melanin-associated protein without direct injury to the eye, i.e. a truly endogenous experimental anterior uveitis[24]. Some infiltration of the choroid occurs but there is no involvement of the retina or breakdown of the blood-retinal barrier. Previously named experimental autoimmune anterior uveitis (EAAU), it is now known as experimental melanin-protein-induced uveitis (EMIU).[25] This model has many similarities in its mode of induction to experimental autoimmune posterior uveitis (EAU) induced by the intradermal injection of retinal antigens (see below), and also to the long list of experimental autoimmune models in other tissues and organs which utilise intradermal injection of autoantigen in adjuvant (e.g., experimental autoimmune encephalomyelitis, experimental autoimmune arthritis, experimental autoimmune thyroiditis, etc.). The nature of the autoantigen in EMIU is under intensive investigation.

POSTERIOR UVEITIS On the notion that sympathetic ophthalmia, occurring several weeks after penetrating ocular injury, is an autoimmune disease affecting the uninjured eye and is caused by sensitisation of the lymphoid tissues to autoantigens released from the injured eye, many attempts have been made to identify autoantigens from the presumptive target tissue, i.e. the uveal tract. However, these were generally unsuccessful. In 1965, Wacker and Lipton[26] demonstrated that a single intradermal inoculation of a crude retinal extract in Freund's complete adjuvant induced a severe panophthalmitis in guinea pigs, experimental autoimmune uveoretinitis (EAU). They went on to show that the major antigen in their extract was a soluble protein of approximately 50kD, which they termed retinal S antigen,[27] later shown to be identical with retinal arrestin (for a review, see ref. 28). Since then, several retinal antigens have been described, including interphotoreceptor retinal binding protein (IRBP), (rhod)-opsin, phosducin and recoverin.[28-30]

Histopathologically, these antigens produce a range of lesions, including inflammatory cell infiltration in the vitreous, microgranulomatous lesions in the retina and especially at the retinal/choroidal interface, retinal vasculitis and in severe acute forms of the disease, extensive retinal detachment and retinal necrosis. The pathology has been fully described in previous publications.[15,31,32]

Generally, the pathology varies with the nature and dose of antigen, the species and strain of animal, the immune status of the animal, and the type of adjuvants used. In Lewis rats, S- and IRBP-induced EAU are usually acute self-limiting but severely retinal-destructive conditions. In the guinea pig[32] and particularly the mouse,[33] the disease may be much less severe but chronic. In this respect, the Lewis rat model resembles the acute stage of Vogt-Koyanagi-Disease in humans[15] while the guinea pig and mouse disease is closer to chronic multifocal choroiditis and the various white dot syndromes seen in man.[34]

EAU is a CD4+ T cell mediated disease in which tissue destruction is mediated by cytokines such as TNFα released from T cells, macrophages and probably NK cells (Dick et al., submitted; Forrester et al., in preparation). CD8+ T cells do not appear to be involved in the active stage of the disease while B cells appear during the late, healing stage and may be immunoprotective.[35]

EAU and models of anterior uveitis such as EMIU are good models for studies of autoimmune disease. In particular, the antigenic epitopes in S- and IRBP-induced EAU have been described[36,37] and considerable research is currently in progress aimed at defining the mechanisms of disease induction.[38] Since EAU has been described in the literature in the most detail, the remainder of this presentation will focus on EAU as a model system for autoimmunity.

Induction of autoimmune responses in the posterior segment

The eye is considered to be an immunologically privileged site, a term intended to denote an altered, impaired or deviated immune response by the tissue when it is immunologically challenged.[39] Previously, this property of the eye was attributed to the lack of lymphatics[40] and hence the lack of immunological reactions was thought simply to be a failure of access to the tissue by inflammatory cells. However, it is clear both experimentally and clinically that immune responses occur endogenously within the eye; so access to the tissue by lymphocytes is unimpaired under certain conditions.

The privileged status of the eye has also been attributed to a lack of antigen presenting cells (APC).[41] However, recent studies of the iris/ciliary body and the choroid have revealed extensive networks of macrophages and dendritic cells.[42–44] Resident tissue macrophages within the eye appear to be relatively long-lived but dendritic cells have a short half life within the eye (Steptoe et al., in preparation), indicating that they may perform a major sentinel role in controlling immunological reactions and peripheral sensitisation. Immune responses within the eye appear to require an intact oculo-splenic axis.[45]

Iris/ciliary body[46] and choroid[47] dendritic cells are fully functional as APC and splenic dendritic cells have the ability to present retinal antigens to sensitised and naive T cells.[48] In this respect, therefore, they are ideally placed to initiate autoimmune reactions at the choroido-retinal interface where they are positioned to encounter any 'leaked' retinal antigenic peptides released from retinal pigment epithelial cells during phagocytosis of photoreceptor tips.[4] In particular, it is easy to envisage release of antigen during a focal non-specific injury to the RPE or the photoreceptors which is then available for uptake by choroidal dendritic cells and presentation to circulating autoreactive T cells, which are present in the circulation of all normal individuals. Each dendritic cell/macrophage focus would than initiate a microgranulomatous lesion which would appear ophthalmoscopically as a 'white dot'.[34] Obviously, the number and size of each lesion would depend on the amount of autoantigen released and the level of successful antigen presentation locally within the tissues.

In posterior uveitis, lesions occur at two sites: namely, the interface between the retina and choroid, and at the retinal vessels. While antigen presentation is likely to occur at the chorio-retinal interface, identification of 'professional' APC within the retina has been more difficult. Dendritic cells, the major APC for induction of *de novo* immune responses,[49] do not occur in the brain or the retina.[50] Aberrant expression of MHC Class II antigens on a range of retinal cell types has been suggested as a means of atypical antigen presentation by retinal endothelial cells and various glial cells but no clear evidence has emerged. The interesting hypothesis that microglial cells, the 'resident macrophages' of the CNS, also function as APC has not been supported by firm evidence (Sedgwick, J., in preparation). Indeed, the opposite is likely to be true. For the present, the most likely candidate APC in the retina is a small population of perivascular resident macrophages which, unlike most resident macrophages, constitutively express high levels of MHC Class II.[51]

Regulation of the immune response in the eye

Since the uveal tract contains extensive networks of professional APCs, it is surprising that autoimmune reactions do not occur more frequently. In part, this is due to the regulatory effects of tissue cells in damping down potential immunological reactions before they commence. Previous studies have shown that ciliary body epithelial cells inhibit lymphocyte proliferation[52] and secrete a variety of 'immunosuppressive' cytokines and mediators, including TGFβ, vasoactive intestinal peptide, prostaglandins, endogenous steroids and αMSH (for a review, see ref. 38). Some of these have been proposed as mediators of immunological privilege within the eye (see above).

In the posterior segment, retinal Muller cells have also been shown to inhibit lymphocyte proliferation in vitro, partly through release of TGFβ.[53] Similarly, microglial cells may be immunoregulatory (see above).

Considerable work has been undertaken on identifying the role of the retinal pigment epithelium (RPE) in regulating the immune response within the posterior segment. Under physiological conditions, the RPE provides an immunosuppressive microenvironment through release of such mediators as PGE_2 and NO, particularly if it comes into contact with small numbers of activated T cells.[54,55] However, the RPE may respond differently in the presence of different cytokine cocktails. When RPE cells are activated by IL-1 and IFNγ, they are themselves induced to release cytokines and chemokines such as IL-8, RANTES and MCP-1[56-58] (Kuppner, unpublished results) which would promote monocyte migration and granuloma formation. In contrast, when RPE cells are exposed to IL-1 and TGFβ they preferentially secrete IL-6 which is the 'damage limitation' cytokine and promotes T cells to secrete IL-4 and IL-13. These concepts have been reviewed in a parallel publication.[38]

Finally, RPE cells may stimulate lymphocyte proliferation in an antigen non-specific manner if the number of T cells becomes sufficiently high, an effect which appears to be mediated by expression of CD59 by the RPE cells (Liversidge et al., submitted).

It is clear from the above that the ocular microenvironment closely regulates immune responses within the eye and that these mechanisms probably account for the older concept of 'immune privilege'. However, when the circumstances change, those same cells which secrete immunosuppressive factors under normal conditions may be induced to participate in and to promote the immune response by aiding recruitment of tissue damaging cells such as macrophages.

Mechanisms of induction of autoimmune response in the eye

As seen from the above, organ-specific autoimmune responses are characterised by (auto)antigen presentation within the target tissue. However, the disease is critically dependent on recruitment of activated antigen-specific T cells to the tissue.[59] In addition, sufficient numbers of antigen-specific T cells, the number of which is as yet unknown, are required to overcome the natural immunosuppressive properties of each tissue, which in the eye are strong (immune privilege). A major question therefore is how are sufficient numbers of autoreactive T cells generated and activated to initiate an immune response?

Since autoreactive T and B cells occur normally in the periphery, albeit at low levels, their deletion during development of self-tolerance is clearly incomplete.[60] Such cells do not induce autoimmune disease in the majority of individuals since they do not come in contact with antigen, either because it is inaccessible (sequestered) due to blood-tissue barriers, or because the dominant epitopes on the antigenic molecules are 'cryptic'. Even if they do come in contact with antigen, they might 'ignore' it unless the conditions are right for the induction of an antigen-specific response.[61]

Recent studies on EAE (for a review, see ref. 62) have revealed that the pathogenicity of normally-circulating autoreactive T cells can be altered by the conditions in which the antigen is detected and/or presented. In a transgenic mouse model of EAE, in which the inserted gene was a functional TCR specific for a dominant epitope on MBP and in which almost all of the T cells expressed the MBP-specific TCR, spontaneous EAE developed at a reproducible frequency (ca. 15%) only when the mice were reared in a non-specific pathogen-free environment. This suggested that concurrent, low grade infection was required to polyclonally activate T cells to a level where sufficient numbers of activated, autoreactive T cells gain access to the target tissue. However, the frequency of spontaneous EAE increased to 100% in a related model of transgenic/knockout mice which, in addition to expressing the MBP/TCR gene, were $RAG^{-}/^{-}$ thereby depleting them of all functional T and B cells except those expressing the MBP/TCR.[63]

These results not only confirm that CD4+ T cells are the only cells necessary to initiate organ-specific autoimmune disease, but also indicate that 'other factors' are active in regulating and/or inhibiting the development of autoimmune disease. Furthermore, the results suggest that amongst these 'other factors' low grade infection may play a part in induction of autoimmune disease.

At least two mechanisms whereby infection with microorganisms may participate in the initiation of autoimmune disease have been proposed: molecular mimicry and polyclonal activation via superantigens.

Molecular mimicry as a concept for the induction of autoimmune diseases has several attractions but is difficult to investigate in human disease, particularly in endogenous posterior uveitis where the link with bacterial infection is tenuous. In acute anterior uveitis (AAU), as in reactive arthritis, the situation is different. As indicated in the introduction, AAU is a recurrent inflammatory disorder which frequently presents in delayed association with a recent infection, commonly an enteritis or UTI.[64] The time interval between the onset of AAU and the initial infection is in the region of a few weeks.

Molecular mimicry is based on the fact that there is homology between known pathogenic peptides from autoantigens and short peptide sequences from microorganisms. This has also been shown for retinal antigens.[65-67] Since many normal individuals have low levels of autoreactive T and B cells (see above), it is conceivable that infection with a microorganism carrying cross-reactive antigenic peptides might induce activation not only of T cells against 'foreign' peptides, but also against self peptides. If sufficient numbers of 'foreign antigen-activated' autoreactive T cells gain access to the eye, uveitis would ensue and would be perpetuated by bystander damage and release of further autoantigen.

Alternatively, sufficient numbers of activated autoreactive T cells could be generated by superantigens which polyclonally activate T cells through non-conventional binding regions on the T cell receptor, thereby 'non-specifically' activating the T cells. Recent studies have indeed shown that a staphylococcal superantigen contains peptide sequences homologous to regions in retinal S antigen (Tighe et al., unpublished).

Both such mechanisms require evidence of restricted T cell activation since any level of antigen specificity will cause monoclonal expansion of T cell populations. Several attempts to provide such evidence have been made in 'classical' autoimmune diseases such as rheumatoid arthritis and multiple sclerosis and also in several experimental models (for a review, see ref. 68). However, at present the evidence is conflicting. In a recent study on posterior uveitis, some degree of oligoclonal T cell restriction was found when only the activated T cell population from the peripheral blood was studied.[69] Interestingly, similar oligoclonality of the T cell population was detected in peripheral blood but not in the joint fluid in patients with rheumatoid arthritis, suggesting that most of the cells in the tissues are non-antigen specific but activated T cells.[70] Further work in this field is continuing.

Conclusion

The aim of this chapter is therefore to demonstrate that models of experimental uveitis have much to offer in the study of autoimmune disease, not only with regard to defining mechanisms of disease induction and identifying relevant antigenic peptides but also in the context of developing therapeutic strategies, since most clinical types of uveitis have their counterpart in an experimental model.

Acknowledgements

The work in this report was supported by the Wellcome Trust and the Guide Dogs for the Blind Association.

References

1 Forrester JV. Uveitis: pathogenesis. Lancet 1991;338:1498-1501.
2 Burmester GR, Daser A, Kamradt T, Krause A, Mitchison NA, Sieper J, Wolf N. Immunology of the reactive arthritides. Ann Rev Immunol 1995;13:229-50.
3 Chan C-C, Palestine MR, Kuwabara T, Nussenblatt RB. Immunohistopathology of ocular sarcoidosis. Arch Ophthalmol 1987;105:1398-1402.

4 Forrester JV. Duke Elder Lecture: New concepts on the role of autoimmunity in the pathogenesis of uveitis. Eye 1992;6:433.
5 Kim MK, Chan C-C, Belfort JR, Farah M, Burnier MP, Nussenblatt RB, Kuwabara T, Palestine AG. Histopathologic features of subretinal fibrosis and uveitis syndrome. Am J Ophthalmol 1987;104:15-23.
6 Ayo C. A toxic ocular reaction. I. New property of Schwartzman toxins. J Immunol 1943;46:113-25.
7 Forrester JV, Worgul BV, Merriam GR. Endotoxin-induced uveitis in the rat. Albrecht von Graefe's Arch klin exp Ophthalmol 1980;213:221-33.
8 Rosenbaum JT, McDevitt HO, Guss RB, Egbert PR. Endotoxin-induced uveitis in rats: a model for human disease. Nature 1980;286:611-3.
9 Hoekzema R, Murray PI, Van Haren MAC, Helle M, Kijlstra A. Analysis of interleukin-6 in endotoxin-induced uveitis. Invest Ophthalmol Vis Sci 1991;32:88-95.
10 Yoshida M, Yoshimura N, Hangai M, Tanihara H, Honda Y. Interleukin-1a, interleukin-1b, and tumour necrosis factor gene expression in endotoxin-induced uveitis. Invest Ophthalmol Vis Sci 1994;35:1107-13.
11 Rosenbaum JT, Samples JR, Hefeneider SH, Howes EL. Ocular inflammatory effects of intravitreal interleukin-1. Arch Ophthalmol 1987;105:1117-20.
12 Rosenbaum JT, Howes EL, Rubin RM, Samples JR. Ocular inflammatory effects of intravitreally injected tumor necrosis factor. Am J Pathol 1988;133:47-53.
13 De Vos AF, Van Haren MAC, Verhagen C, Hoekzema R, Kijlstra A. Systemic anti-tumour necrosis factor antibody treatment exacerbates endotoxin-induced uveitis in the rat. Exp Eye Res 1995;61:667-76.
14 Trentham DE, Dynesius-Trentham R. Collagen-induced arthritis. In: Mechanisms and Models in Rheumatoid Arthritis, 1995:447-56.
15 Forrester JV, Liversidge JM, Dua HS, Towler HM, McMenamin PG. Comparison of clinical and experimental uveitis. Curr Eye Res 1990;9(Suppl):75-84.
16 Doran MC, Goodstone NJ, Hobbs RN, Ashton BA. Cellular immunity to cartilage link protein in patients with inflammatory arthritis and non-arthritic controls. Ann Rheum Dis 1995;54(6):466-70.
17 Smith RE, Jensen A, Silverstein AM. Antibody formation by single cells during experimental immunogenic uveitis. Invest Ophthalmol 1969;8:373.
18 Pribnow JF, Hall JM. Studies on the intravitreal initiation of the immune response. Invest Ophthalmol 1970;9:639.
19 Liu SH, Predergast RA, Silverstein AM. The role of lymphokines in immunogenic uveitis. Invest Ophthalmol Vis Sci 1983;24:361-7.
20 Van der Voet JCM, Liem A, Otto AJ, Kijlstra A. Intraocular antibody synthesis during experimental uveitis. Invest Ophthalmol Vis Sci 1989;30:316-22.
21 Marak GE, Font RL, Alepa FP. Arthus-type panophthalmitis in rats sensitised to heterologous lens protein. Ophthalm Res 1977;9:162-70.
22 Takagi T, Nakajima A, Yamaki K, Sakuragi S. Clinical course, immune response and histopathology of a simplified model of experimental lens-induced endophthalmitis in rats. Ocular Immunol Inflam 1995;3:157-62.
23 Wucherpfennig KW, Stromiger JL. Molecular mimicry in T cell-mediated autoimmunity: viral peptides activate human T cell clones specific for myelin

basic protein. Cell 1995;80:695-705.
24 Broekhuyse RM, Kuhlmann ED, Winkens HJ, Van Vugt AHM. Experimental autoimmune anterior uveitis (EAAU), a new form of experimental uveitis. I. Induction by a detergent-insoluble, intrinsic protein fraction of the retinal pigment epithelium. Exp Eye Res 1991;52:465-74.
25 Broekhuyse RM, Kuhlmann ED, Winkens HJ. Experimental autoimmune anterior uveitis(EAAU). III. Induction by immunisation with purified uveal and skin melanins. Exp Eye Res 1993;56:401-11.
26 Wacker WB, Lipton MM. Experimental allergic uveitis: homologous retina as uveitogenic antigen. Nature 1965;206:253-8.
27 Wacker WB, Kalsow CM, Yankeelov JA, Organisciack DT. Experimental allergic uveitis. Isolation, characterisation and localisation of a soluble uveitopathogenic antigen from bovine retina. J Immunol 1977;119:1949-58.
28 Gery I. Retina specific antigens and the immunopathologic process they provoke. Progr Retinal Res 1986;5:75-109.
29 Gery I, Chanaud NP, Anglade E. Recoverin is highly uveitogenic in Lewis rats. Invest Ophthalmol Vis Sci 1994;35:3342-5.
30 Dua HS, Lee RH, Lolley RN, Barrett JA, Abrahams M, Forrester JV, Donoso LA. Induction of experimental autoimmune uveitis by the retinal photoreceptor cell protein, phosducin. Curr Eye Res 1992;11(Suppl.):107-12.
31 McMenamin PG, Broekhuyse RM, Forrester JV. Ultrastructural pathology of experimental autoimmune uveitis: a review. Micron 1992;24:521-46.
32 Forrester JV, Borthwick GM, McMenamin PG. Ultrastructural pathology of S-antigen uveoretinitis. Invest Ophthalmol Vis Sci 1985;26:1281-92.
33 Caspi RR, Roberge FG, Chan C-C, Wiggert B, Chader GJ. A new model of autoimmune disease. Experimental autoimmune uveoretinitis induced in mice with two different retinal antigens. J Immunol 1988;140:1490-5.
34 Ben Ezra D, Forrester JV. Fundal white dots: the spectrum of a similar pathological process. Br J Ophthalmol 1995;79:856-60.
35 Liversidge JM, Forrester JV. Experimental allergic uveitis (EAU): immunophenotypic analysis of the inflammatory cells in chorioretinal lesions. Curr Eye Res 1988;7:1231-41.
36 Sanui H, Redmond TM, Kotake S, Wiggert B, Hu L-H, Margalit H, Berzoksky JA, Chader GJ, Gery I. Identification of an immunodominant and highly immunopathogenic determinant in the retinal interphotoreceptor retinoid-binding protein (IRBP). J Exp Med 1989;169:1947-60.
37 Donoso LA, Merryman CF, Sery TW, Shinohara T, Dietzschold B, Smith A, Kalsow C. S-antigen: characterisation of a pathogenic epitope which mediates experimental autoimmune uveitis and pinealitis in Lewis rats. Curr Eye Res 1987;6:1151-9.
38 Forrester JV, Lumsden L, Liversidge J, Kuppner M, Mesri M. Immunoregulation of uveretinal inflammation. Progr Retinal Eye Res 1995;14:393-412.
39 Streilein JW, Willbanks GA, Cousins SW. Immunoregulatory mechanisms of the eye. J Neuroimmunol 1992;39:185-200.
40 Tompsett E, al-Hanna D, Wakefield D. Immunological privilege in the eye. Curr Eye Res 1990;12:1141-5.
41 Streilein JW, Willbanks GA, Taylor A, Cousins S. Eye-derived cytokines and

the immunosuppressive micro-environment: a review. Curr Eye Res 1992;11S:41-7.

42 McMenamin PG, Crewe J, Morrison S, Holt PG: Immunomorphological studies of macrophages and MHC Class II-positive dendritic cells in the iris and ciliary body of the rat, mouse and human eye. Invest Ophthalmol Vis Sci 1994;35:3234-50.

43 McMenamin PG, Holthouse I, Holt PG. Class II MHC (Ia) antigen-bearing dendritic cells within the iris and ciliary body of the rat eye: Distribution, phenotype, ontogeny, and relation to retinal microglia. Immunology 1992;77:385-93.

44 Forrester JV, McMenamin PG, Holthouse I, Lumsden L, Liversidge J. Localisation and characterisation of major histocompatibility complex Class II-positive cells in the posterior segment of the eye: implications for the induction of autoimmune uveoretinitis. Invest Ophthalmol Vis Sci 1994;35:64-77.

45 Streilein JW, Niederkorn JY. Induction of anterior chamber-associated immune deviation requires an intact functional spleen. J Exp Med 1981;153:1058-67.

46 Steptoe RJ, Holt PG, McMenamin PG. Functional studies of major histocompatibility class II-positive dendritic cells and resident macrophages isolated from the rat iris. Immunology 1995;58:630-8.

47 Choudhury A, Pakalnis VA, Bowers WE. Characterisation and functional activity of dendritic cells from rat choroid. Exp Eye Res 1994; in press.

48 Forrester JV, McMenamin PG, Liversidge J, Lumsden L. Dendritic cells and 'dendritic' macrophages in the uveal tract. New York/London: Plenum Press, 1993.

49 Steinman RM. The dendritic cell system and its rôle in immunogenicity. Annu Rev Immunol 1991;9:271-96.

50 Sedgwick JD, Schwender S, Imrich H, Dorries R, Butcher GW, Ter Meulen V. Isolation and characterisation of resident microglial cells from the normal and inflamed central nervous system. Proc Natl Acad Sci USA 1991;88:7438.

51 Dick AD, Ford AL, Forrester JV, Sedgwick J. Flow cytometric identification of a minority population of MHC Class II positive cells in the normal rat retina distinct from $CD45^{low}CD11b/c+CD4^{low}$ parenchymal microglia. Br J Ophthalmol 1995;79:834-40.

52 Helbig H, Gurley RC, Palestine AG, Nussenblatt RB, Caspi RR. Dual effect of ciliary body cells on T lymphocyte proliferation. Eur J Immunol 1990;20:2457-63.

53 Caspi RR, Roberge RG, Nussenblatt RB. Organ resident non-lymphoid cells suppress proliferaton of autoimmune T-helper lymphocytes. Science 1987;237:1029-32.

54 Liversidge JM, McKay D, Mullen G, Forrester JV. Retinal pigment epithelial cells modulate lymphocyte function at the blood-retina barrier by autocrine PGE2 and membrane bound mechanisms. Cell Immunol 1993;149:315-30.

55 Liversidge J, Grabowski P, Ralston S, Benjamin N, Forrester JV. Rat retinal pigment epithelial cells express an inducible form of nitric oxide synthase and produce nitric oxide in response to inflammatory cytokines and activated T cells. Immunology 1994;in press.

56 Kuppner MC, McKillop-Smith S, Forrester JV. Transforming growth factor b and IL-1 act in synergy to enhance IL6 and IL8 mRNA levels and IL6 production by human retinal pigment epithelial cells. Immunology 1994; in press.
57 Elner SG, Strieter RM, Elner VM, Rollins BJ, Del Monte MA, Kunkel SL. Monocyte chemotactic protein gene expression by cytokine-treated human retinal pigment epithelial cells. Lab Invest 1991;64:819-25.
58 Elner VM, Streitter RM, Elner SG, Lindley L, Kunkel SL. Neutrophil chemotactic factor(IL8) gene expression by cytokine treated retinal pigment epithelial cells. Am J Pathol 1990;136:745.
59 Dick AD. Retinal antigen-specific T cells mediate experimental autoimmune uveoretinitis (EAU) in the PVG rat. Ocular Immunol Inflam 1995;3:261-70.
60 Von Boehmer HL. Developmental biology of T cells in T-cell receptor transgenic mice. Annu Rev Immunol 1990;8:531.
61 Elson CJ, Barker RN, Thompson SJ, Williams NA. Immunologically ignorant autoreactive T cells, epitope spreading and repertoire limitation. Immunol Today 1995;16:71-5.
62 Hayday A. Is antigen-specific immunosuppression now unsuppressed? Curr Biol 1995;5:47-50.
63 Lafille J, Nagashima K, Katsuki M, Tonegawa S. High incidence of spontaneous autoimmune encephalomyelitis in immunodeficient anti-myelin basic protein T-cell receptor transgenic mice. Cell 1994;78:399-408.
64 Rothova A. 865 consecutive uveitis patients assessed. Br J Ophthalmol 1992;76:137-41.
65 Shinohara T, Donoso L, Tsuda M, Yamaki K, Singh VK. S-antigen: structure, function and experimental autoimmune uveitis (EAU). Progr Retinal Res 1988;8:51-66.
66 Singh VK, Yamaki K, Donoso LA, Shinohara T. Yeast histone H3-induced experimental autoimmune uveoretinitis. J Immunol 1989;142:1521.
67 Singh VK, Yamaki K, Donoso LA, Shinohara T. Sequence homology between yeast histone H3 and uveitopathogenic site of S-antigen: Lymphocyte cross-reaction and adoptive transfer of the disease. Cell Immunol 1989;119:211-21.
68 Heber-Katz E, Acha-Orbea H. The V-region hypothesis: evidence from autoimmune encephalomyelitis. Immunol Today 1989;10:164-9.
69 Tighe P, Forrester JV, Liversidge J, Sewell HF. Peripheral CD25 positive T lymphocytes with a biased T cell receptor Vb usage in autoimmune endogenous posterior uveitis. J Clin Pathol: Molec Pathol 1995;48:M46-50.
70 Goronzy JJ, Bartz-Bazzanella P, Weining H, Jendro MC, Walser-Kuntz DR, Weyand CM. Dominant clonotypes in the repertoire of peripheral CD4+ T cells in rheumatoid arthritis. J Clin Invest 1994;94:2086.

MANFRED ZIERHUT, RAINER STIEMER, BERNHARD KREUTZER AND HANS-JÜRGEN THIEL

Clinical uveitis and autoimmunity

Introduction

Inflammation of the uvea, which comprises ciliary body, iris and choroid, is called "uveitis". Up to now there are approximately 100 known etiologies leading to uveitis.[1] For most of these entities the underlying immunomechanism remains unclear.

Figure 1 illustrates the factors which are important in our understanding of the mechanisms of uveitis. Besides microbial antigens, autoantigens have been suggested as playing a major role in uveitis. The application of autoantigens can produce experimental autoimmune uveoretinitis (EAU). With this model it has been possible to study the immune response of the eye in a T-cell dependent manner. The phenomenon of anterior chamber associated immune deviation (ACAID) (see p. 87-90 in this book), antigen-presenting cells and genetic factors influence the response to both EAU and clinical uveitis.

This chapter summarizes the mechanisms of well established clinical types of uveitis and compares them with results from studies of the EAU model.

Definition of uveitis

As recommended by the International Uveitis Study Group,[2] the following pathologic-anatomical nomenclature will be used:

anterior uveitis: inflammation of the iris, sometimes also with inflammation of the vitreous,

intermediate uveitis: inflammation of the vitreous without choroidal changes, but sometimes with mild circumscribed anterior chamber inflammation,

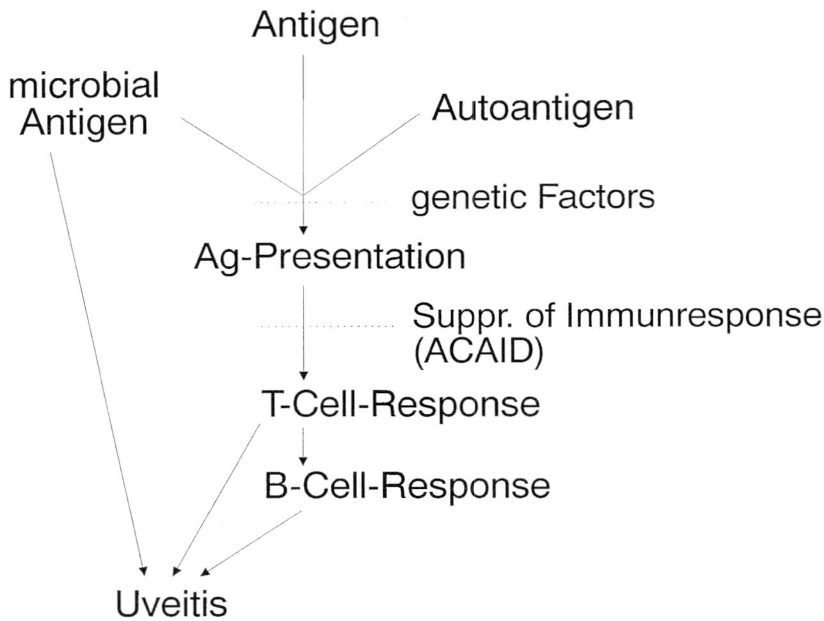

FIGURE 1. Factors playing a role in the mechanism of uveitis.

posterior uveitis: inflammation of the choroid or retina with secondary choroiditis.

The incidence of uveitis is approximately 20/100,000 per year, leading to 10% of the blind registrations in the civilized world.

Mechanisms of uveitis

Uveitis can be caused by various antigens. Only a few microorganisms are known to induce uveitis, which can be initiated after exogenous but more commonly after endogenous infections. This is well described for lues, borreliosis, tuberculosis and other microorganisms. Depending on the kind of infection, this may directly lead to uveitis if the cytopathic effect dominates. A second way to initiate inflammation is by stimulation of the immune system by the microorganism, leading to an immunopathological effect. After antigen uptake and presentation a T-cell response may induce uveitis.

A third way can be demonstrated in the model of EAU when rats or mice are immunized by well- characterized autoantigens like retinal S-antigen or IRBP (interphotoreceptor retinoid binding protein). In this model, the autoantigen induces a T-cell response. This immune reaction can be studied and comparison with human uveitis has disclosed various similarities, which means that some clinical uveitis entities have characteristics typical for an autoimmune disease.

Autoimmunity and uveitis

There are a few characteristics of autoimmune diseases. These often include a dominant role for autoreactive T cells. The EAU model has disclosed that this disease can be transferred by T cells. The role of B cells in EAU seems not to be that important. On the other hand, production of autoantibodies (e.g. anti-retinal-S-ag) can theoretically have an impact on the protection from retinal and uveal tissue damage. As in most autoimmune disorders, there is a genetic role for the development of uveitis but this depends strongly on the type of uveitis. A hereditary component, which is found in only a few autoimmune disorders, is quite rare in uveitis as well. These aspects will be discussed in more detail in the following sections.

Immunogenetics

HLA-ASSOCIATIONS A variety of autoimmune disorders show a high association with HLA-antigens. The reason for this is not very well understood and may depend on the HLA-type and the disorder.

HLA-B27 The association between ankylosing spondylitis (AS) and HLA-B27 was the first to be described.[3] While approximately 90% of AS patients express this antigen, 43-60% of patients with acute anterior uveitis (AAU) are HLA-B27 positive. Patients with both disorders are nearly all HLA-B27 positive.

Previous investigations have shown that HLA-B27 positive AAU patients may differ clinically from HLA-B27 negative AAU patients, having more and stronger recurrences, but the nine-year prognosis remains the same.[4]

How HLA-B27 leads to AS and AAU is unknown. Various hypotheses have been published discussing a dominant role for gram-negative bacteria (Yersinia, Klebsiella, Chlamydia and others). Recently, molecular mimicry between HLA-B27 and retinal S-antigen has been observed.[5] While T cells cross-react between a peptide sequence from HLA-B27 and a peptide from retinal S-antigen, it was also possible to induce oral tolerance with the HLA-B27 peptide in the EAU model. These results do not explain why HLA-B27 associated disease typically leads exclusively to anterior uveitis while retinal S-antigen in high concentrations is found in the retina only, which is completely spared in the HLA-B27 associated disease.

Remarkably, in AAU associated with HLA-B27, only one eye is inflamed while the other eye may display active inflammation later. A role for local factors such as TGF-ß has been discussed. This leads to the theory that the anterior chamber is always balancing on the razor´s edge between a level of phagocytosing activities leading to normal cleaning of the anterior chamber and a level of hyperactive cleaning leading to inflammation.[6] Additionally, environmental factors may be important in HLA-B27 associated disorders.

HLA-B51 This antigen has been found to be associated with Behçet´s disease. Especially in countries with high prevalence like the Mediterranean countries and Japan, but also in Germany, there is a high association between HLA-B51 and Behçet´s disease (Table 1).[7] The reason for this association is unclear at the moment.

TABLE 1 Survey of HLA-B5/B51 associations among studies from different countries.

Country	Behcet´s disease	Healthy controls
Japan	57%	12%
Israel	71%	13%
Italy	76%	8%
Turkey	62%	12%
France	65%	9%
USA	12%	10%
Spain	52%	20%
England	25%	6%
Germany[7]		
Mediterraneans	54%	12%
Germans	67%	6%

HLA-A29 In 80-96% of the cases, birdshot retinopathy is associated with HLA-A29, which is found in only up to 7% of the healthy population.[8] Using one-dimensional isoelectric focusing techniques, two subtypes are separable. While type 1 is found in up to 10% of patients (7-10% in healthy controls), type 2 is detectable in 90-100% of birdshot patients. But type 2 is also the one regularly found in the control group (90-93%).[9] The HLA-A29 peptide binding motif has recently been identified.[10]

HLA-DR4/DQw1 Vogt-Koyanagi-Harada (VKH) disease has been shown to be associated with HLA-DR4 in 75-88 % of the cases (normal population 23-32 %) and with HLA-DRw53 in 80% (normal population 15%). HLA-DR4 includes at least 12 gene segments, DRB1*0401 to DRB1*0412. Clinically, the VKH-syndrome may appear in two forms, a prolonged and a non-prolonged type. In a previous study, the prolonged type has been associated with DRB1*0405 or 0410 in 100% of the cases, the non-prolonged type only in 61%.[11] This leads to the conclusion that the gene variant may determine the severity of VKH. It has been found that HLA-DQB1*0604 is negatively associated with the VKH-syndrome.[12]

NON-HLA-ASSOCIATED HEREDITARY UVEITIS Uveitis is not a hereditary autoimmune disorder. Despite this there are a few reports in the literature on hereditary uveitis. Besides HLA-B27 associated anterior uveitis and three reports of pars planitis, one report of Behçet´s disease and one report of Fuchs' heterochromia, birdshot retinopathy has also been found in monozygous twins. In 1985, Blau et al.[13] reported a form of hereditary uveitis consisting of anterior uveitis, symmetric non-erosive arthritis and a skin rash. It is transmitted in an autosomal-dominant fashion. Figure 2 shows a family, seen at the University Eye Clinic in Tübingen, with 6 affected members over 4 generations. HLA-class-I and class-II typing disclosed no HLA-associations. Additionally, all members were HLA-B27 negative, whether or not they had uveitis. This family clearly demonstrated anticipation (earlier onset and worsening of symptoms in succeeding generations).

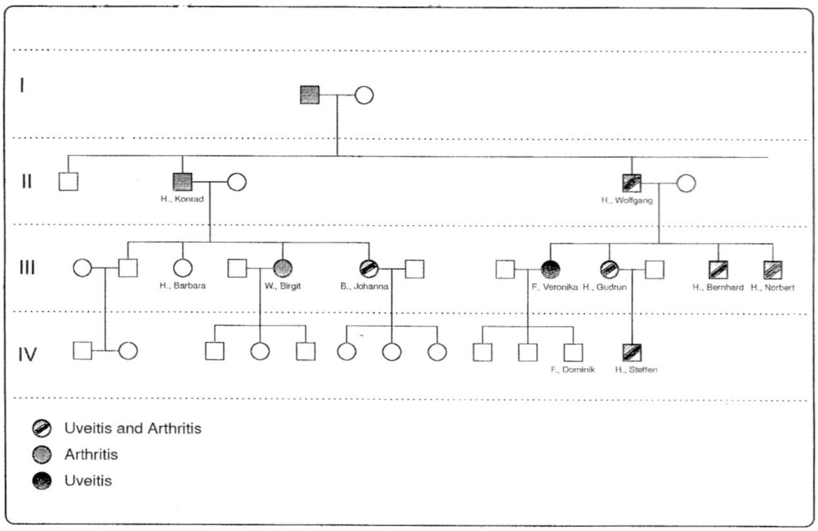

FIGURE 2. Autosomal dominant heredity of anterior uveitis in the Blau Syndrome.

Immunopathology of various uveitis types

The study of cell populations of uveal tissues using immunohistological techniques has been performed only in some classical forms of uveitis. The drawback of these studies is that the examined eyes were mostly enucleated in a late stage of disease. Therefore, the results must be interpreted with caution regarding the immunopathological mechanisms of the disease.

INTERMEDIATE UVEITIS Intermediate uveitis is characterized by inflammation of the vitreous with minor anterior chamber involvement and absence of choroidal lesions. Often, additional vasculitis of venous vessels is detectable, rarely neovascularization and exudative retinal detachment.

The immunohistopathology is characterized by CD4+ T cells around retinal vessels and also in snowballs, peripherally located conglomerates of lymphocytes.[14] Snowballs also stain positive with antibody to glial fibrillary acidic protein, suggesting a role for Müller cells in this disorder. So far, no B cells have been found in intermediate uveitis.

SYMPATHETIC OPHTHALMIA Sympathetic ophthalmia (SO) is a bilateral panuveitis induced by penetrating injuries or (less often) intraocular surgery. The etiology is unknown, but an autoimmune reaction against a retinal antigen after microbial infection has been discussed for years.

Histologically, SO is characterized by a granulomatous chorioretinitis. These granulomas are called Fuchs-Dalen nodules, which are located at the retinal pigment epithelium and consist of HLA-class-II antigen-expressing epitheloid cells. While mainly CD4+ T cells are detectable in the early stage of the disease, later a predominance of CD8+ T cells is found.[15] B cells seem to play a minor role in SO.

VOGT-KOYANAGI-HARADA SYNDROME The Vogt-Koyanagi-Harada (VKH) syndrome is a systemic disorder leading to panuveitis, vitiligo, poliosis and inner-ear hearing impairment. The etiology is unknown. A systemic autoimmune reaction is suggested, directed against an autoantigen common to all affected organs.

Histologically, VKH resembles SO with Fuchs-Dalen granulomas in the uvea. As in SO, CD4+ T cells dominate the early stage of inflammation.[16,17] Macrophages which are HLA-class-II positive are numerous, while B cells are rarely detectable in the retina. In the aqueous humor, the ratio of CD4/CD8 T cells has been found to be 52%/32%,[18] similar to the HLA-class-II + cells which are seen in the aqueous humor in 52% and in the CSF in 31%.[18] While IL-6 is elevated in the aqueous humor but not in the blood, levels of IL-2 and soluble IL-2 receptor are within normal limits in both serum and aqueous humour.[18]

BEHÇET'S DISEASE Behçet's disease is a multisystem disorder, characterized by oral and genital aphthosis, uveitis, thrombosis, arthritis, gastrointestinal problems and CNS involvement.

Histology of uveal biopsy discloses a perivascular infiltration of polymorphous nucleocytes and lymphocytes. By means of immunohistology, CD4+ T cells bearing large numbers of IL-2 receptors have been identified.[19] Serum IL-2 levels have also been detected.[20]

SARCOID Sarcoid is a systemic disorder which may lead to involvement of the eye, the skin, the joints and the CNS, but also various other systems. The etiology remains unknown.

All types of uveitis can be associated with sarcoid. Anterior uveitis is found in 22-85% of ocular sarcoid, intermediate uveitis in 10%, and posterior uveitis in 14-43%. The clinical picture can include nodules in the iris and/or choroid, retinal vasculitis, pars plana infiltration, candle wax drippings, and also neovascularization. Histology of sarcoid nodules in the iris or choroid discloses non-caseating granulomas consisting of whorls of epitheloid cells surrounding multinucleated giant cells. Mononuclear cells, fibroblasts, and lymphocytes may be found at the periphery of these nodules.

Immunohistopathological examination has been reported in only very few patients. Chan et al.[21] found more CD4 than CD8 cells (10:1) in the choroid with epitheloid cell granulomas, and the CD8 cells were grouped ringlike around the granulomas. At the periphery of the granulomas B cells were sometimes detectable. Granuloma-forming cells were positive for HLA-class II antigen, Il-2 receptor and IFN-γ. In the serum of sarcoid-induced uveitis patients, significantly elevated titers against ICAM-1[22] and elevated titers against soluble IL-2 receptor and Il-2 receptor have been found. B cells are elevated, and Berthoud et al.[23] found elevated anti-HSV-antibodies in these patients, which may be the result of a polyclonal B-cell activation.

Similarities and differences between EAU and clinical uveitis

Clinical and experimental uveitis exhibit some similarities but also some differences. Both disorders are characterized by similar pathology. This includes cells in the anterior chamber, focal lesions, vasculitis, subretinal neovascularization and exudative retinal detachment. In a dose-dependent way EAU can result in anterior uveitis, intermediate uveitis or posterior uveitis (Table 2).

There are also some differences between experimental and clinical uveitis. Whereas in EAU, rats and mice, for example, develop more

TABLE 2 Dependence between inoculated retinal S-antigen dosis and uveitis form in the guinea pig EAU model.[24]

50 µg	panophthalmitis
25 µg	endophthalmitis
5-10 µg	granulomatous posterior uveitis
1 µg	non-granulomatous posterior uveitis
mildest form:	intermediate uveitis

infiltration of the retina, uveitis in humans and monkeys is more localized to the choroid. Experimental uveitis has a self-limiting course, clinical uveitis frequently recurs.

In the following paragraphs, some important immunological parameters and immune responses in EAU will be compared with those in clinical uveitis.

T CELLS It has been known for years that EAU is a T-cell dependent disease, since nude mice are not able to develop uveitis.[25] In addition, uveitogenic T-cell lines are able to transfer the disease to naive rats.[26] These cells have been characterized as Ia-, IL-2 receptor+ CD4+ cells. In a later stage of the disease, CD8+ suppressor/cytotoxic T-cells accumulate in the retina.[27] For EAU it has been shown that mast cells in the choroid degranulate just before the influx of T cells into the eye begins.[28]

In clinical uveitis, similar findings have been published for ocular tissue and ocular fluids, with the exception of Fuchs' heterochromic cyclitis in which B cells seem to dominate because high concentrations of IgG have been found in the anterior chamber,[29] together with high levels of interleukin-6.[30]

Deschenes et al.[31] demonstrated an increased number of IL-2 receptor + T cells in the peripheral blood of uveitis patients. A positive correlation was seen between the degree of lymphocyte activation and clinical uveitis activity.

With regard to the role of mast cells in human uveitis, it seems that they do not play such a dominant role as in EAU.

T cells may reach the retina by adherence in high endothelial venules (HEV), which are normally not found in the retina of humans or mice.[32] After activation by cytokines which may be released by Müller cells or the very first invading T cells, HEV´s are detectable around day 10 in the retina in the murine IRBP-model (manuscript in preparation). Blocking of ICAM-1 with monoclonal antibodies was not able to prevent HEV activation, but did prevent EAU (submitted for publication).

TCR The preferential usage of one T-cell receptor (TCR) may indicate an immune reaction directed against one autoantigen. For EAU induced by retinal S-antigen, a preferential use of Vß8.2 has been shown in the rat model,[33] while in the mouse model Vß2, Vß12 and Vß15 are mostly found on invading T cells.[34] In the early stages of IRBP-induced EAU, T cells carry both Vß8.2 and Vß8.3.[33] Immunization with Vß8.2 fragments was not successful in EAU,[35] but did prevent disease in the model of experimental autoimmune encephalitis.[36]

So far, no preferential usage of any T-cell receptor has been published in any human study, only a relative abundance of TCR Vγ/δ T cells in the lacrimal gland in patients with ocular sarcoid.[37,38] The main reason for this may be that the onset of the uveitis has always been missed when a possible preferential usage of one TCR exists.

CELLULAR IMMUNE RESPONSE TO RETINAL ANTIGENS T cells of uveitis patients generally respond more strongly to IRBP than to retinal S-antigen, but anti-IRBP Ab titers seem to be less predictive of active disease than anti-S-antigen Ab titers. The stimulation index is higher in posterior uveitis than in intermediate uveitis, where it is still higher than in controls.[39]

In Behçet´s disease, T cells are directed against both autoantigens and fragments of autoantigens.[40] This may reflect the extent of the ocular damage and may possibly be used as a marker for activity.

AUTOANTIBODIES IN UVEITIS Although B cells seem to be of little importance in the etiology of most uveitis forms, including EAU, the autoantibodies (AAb) produced could be of diagnostic relevance. It is possible to differentiate autoantibodies that are directed against specific ocular antigens from those that are not specific for ocular antigens.

AAb against ocular antigens The anti-retinal S-antigen response has been studied, but with inconsistent results. In 1985, Chan et al.[41] found these AAb´s in the serum of VKH and SO patients, but not in patients with Behçet's disease. In 1991, Gregerson et al.[42] detected these AAb´s in iritis, toxoplasmosis, sarcoid and also Behçet's disease. La Hay et al.[43] found higher titers in Fuchs' heterochromic cyclitis patients, but also in the serum of patients with retinitis pigmentosa, diabetic retinopathy and retinal detachment.

Doeckes et al.[44] in 1987 and Forrester et al.[45] in 1989 were unable to identify higher titers in various uveitis groups, and Uusitalo et al.[46] did not find them in uveitis of children.

Today, it is not clear whether anti-retinal S-antigen Ab´s are destructive or protective for the ocular tissue. They may be induced by polyclonal activation. Therefore, longitudinal studies, analysis of Ab-subclasses and of selected epitopes of ocular autoantigens are necessary.

AAb against non-ocular tissues Only a limited number of studies have been done regarding non-ocular autoantibody formation in uveitis patients. In the first studies,[47-49] investigating antibodies against smooth muscle cells, parietal cells, reticulin, DS-DNA and mitochondria, only anti-nuclear antibodies (ANA) were detected in 77% of patients with juvenile arthritis. Our group extended these studies, screening for elevated ab titers against 14 antigens, and we disclosed anti-endothelial antibodies (AEA) in 33% of patients with posterior uveitis, but only in 18% with anterior uveitis.[50] In contrast to this we found anti-sarcolemmal ab´s in 59% of patients with anterior uveitis, but only in 8% of patients with posterior uveitis.

All autoantibodies are detectable in approximately 5% of healthy controls. Patients with intermediate uveitis do not show higher titers than controls.

In the future, detection of AAb patterns may be helpful in the diagnosis of subgroups of uveitis.

Conclusion

Clinical uveitis demonstrates features of classical autoimmune disease. These are the importance of a CD4+ T-cell reaction, the HLA-class II expression on organ-specific cells, HLA-association of some uveitis types and the response to immunosuppressive treatment.

Besides the similarities there are also important differences between experimental and clinical uveitis. While EAU is a self-limiting disease, clinical uveitis most often has a chronic course. There is a preferential usage of the TCR in EAU, but not in human uveitis. The future will yield more information about the initiating antigens or autoantigens in uveitis.

References

1 Zierhut M. Uveitis.Vol.1. Differential Diagnosis. Buren, The Netherlands: Aeolus Press, 1995.

2 Bloch-Michel E, Nussenblatt RB. International uveitis study group recommen-

dations for the evaluation of intraocular inflammatory disease. Am J Ophthalmol 1987;103:234-5.
3 Brewerton DA, Caffrey M, Nicholls A, Walters D, James DC. Anterior uveitis and HL-A 27. Lancet 1973;2:994-6.
4 Linssen A. Acute anterior uveitis, HLA-B27 and the spondylarthropathies. Fifteen years experience in the Netherlands. In: Zierhut M, Saal JG, Thiel HJ, editors. Immunology of the joint and the eye. Buren, The Netherlands: Aeolus Press, 1996:159-70.
5 Wildner G, Thurau SR. Cross-reactivity between an HLA-B27-derived peptide and a retinal autoantigen peptide: a clue to major histocompatibility complex association with autoimmune disease. Eur J Immunol 1994;24:2579-85.
6 Feltkamp TEW. Pathomechanisms in HLA-B27 related diseases. In: Zierhut M, Saal JG, Thiel HJ, editors. Immunology of the joint and the eye. Buren, The Netherlands: Aeolus Press, 1996:123-35.
7 Zierhut M, Saal J, Pleyer U, Kötter I, Dürk H, Fierlbeck G. Behcet´s disease: epidemiology and eye manifestations in German and Mediterranean patients. German J Ophthalmol 1995;4:246-51.
8 Baarsma GS, Priem HA, Kijlstra A. Association of birdshot retinochoroidopathy and HLA-A29 antigen. Curr Eye Res 1990;9 (suppl):63-8.
9 LeHoang P, Ozdemir N, Benhamou Atabary T, Edelson C, Betuel H, Semiglia R, Cohen JHM. HLA-A29.2 subtype associated with birdshot retinochoroidopathy. Am J Ophthalmol 1992;113:33-5.
10 Toubert A, Boisgérault F, Khalil I, Tieng V, Connan F, Tabary T, Choppin J, Charron D. HLA-A29 peptide binding motif. Int Congr Immunol, San Francisco 1995;705 (abstract).
11 Islam SMM, Numaga J, Matsuki K, Fujino Y, Maeda H, Masuda K. Influence of HLA-DRB1 gene variation on the clinical course of Vogt-Koyanagi-Harada disease. Invest Ophthalmol Vis Sci 1994;35:752-6.
12 Islam SMM, Numaga J, Fujino Y, Hirata R, Matsuki K, Maeda H, Masuda K. HLA class II genes in Vogt-Koyanagi-Harada disease. Invest Ophthalmol Vis Sci 1994;35:3890-6.
13 Blau EB. Familial granulomatous arthritis, iritis, and rash. J Pediat 1985;5:689-93.
14 Wetzig RP, Chan CC, Nussenblatt RB, Palestine AG, Mazur DO, Mittal KK. Clinical and immunopathological studies of pars planitis in a family. Br J Ophthalmol 1988;72:5-10.
15 Chan CC, Benezra D, Rodrigues MM, Palestine AG, Hsu SM, Murphree AL, Nussenblatt RB. Immunohistochemistry and electron microscopy of choroidal infiltrates and Dalen-Fuchs nodules in sympathetic ophthalmia. Ophthalmology 1985;92:580-90.
16 Chan CC, Palestine AG, Kuwabara T, Nussenblatt RB. Immunopathologic study of Vogt-Koyanagi-Harada syndrome. Am J Ophthalmol 1988;105:607-11.
17 Kahn M, Pepose JS, Green WR, Miller J, Foos RY. Immunocytologic findings in a case of Vogt-Koyanagi-Harada syndrome. Ophthalomology 1993;100: 1191-8.
18 Norose K, Yano A, Wang X, Tokushima T, Umihira J, Seki A, Nohara M,

Segawa K. Dominance of activated T cells and interleukin-6 in aqueous humor in Vogt-Koyanagi-Harada disease. Invest Ophthalmol Vis Sci 1994;35:33-9.
19 Charteris DG, Champ C, Rosenthal AR, Lightman SL. Behçet's disease: activated T lymphocytes in retinal perivasculitis. Br J Ophthalmol 1992;76:499-501.
20 Ben Ezra D, Maftzir G, Kalichman I, Barak V. Serum levels of interleukin-2 receptor in ocular Behçet´s disease. Am J Ophthalmol 1993;115:26-30.
21 Chan CC, Wetzig RP, Palestine AG, Kuwabara, T, Nussenblatt RB. Immunohistopathology of ocular sarcoidosis: report of a case and discussion of immunopathogenesis. Arch Ophthalmol 1987;105:1398-1402.
22 Arocker-Mettinger EA, Steurer-Georgiew L, Steurer M, Huber-Spitzy V, Asenbauer TT, Steinkogler FJ, Hoelzl E, Kuchar A, Grabner G. Circulating ICAM-1 levels are elevated in the serum of uveitis patients. In: Dernouchamps JP, Verougstraete C, Caspers-Velu L, Tassignon MJ, editors. Recent advances in uveitis. Amsterdam: Kugler Publ, 1993:95-101.
23 Berthoud JF, Keller A, Guex-Crosier Y, Herbort CP. Polyclonal rise of antiherpes antibodies in serum: a diagnostic aid rather than a diagnostic trap in sarcoidosis uveitis. Invest Ophthalmol Vis Sci 1994;35(4):2099.
24 Rao NA, Wacker WB, Marak Jr GE. Experimental allergic uveitis. Clinicopathologic features associated with varying doses of S-antigen. Arch Ophthalmol 1979; 97:1954-8.
25 Salinas-Carmona MC, Nussenblatt RB, Gery I. Experimental autoimmune uveitis in the athymic nude rat. Eur J Immunol 1982;25:481-4
26 Rozenszajn LA, Muellenberg-Coulombre C, Gery I, El-Saied M, Kuwabara T, Mochizuki M, Lando Z, Nussenblatt RB. Induction of experimental autoimmune uveoretinitis (EAU) in rats by T cell lines. Immunology 1986;57: 559-65.
27 Chan CC, Mochizuki M, Palestine AG, Ben Ezra D, Gery I, Nussenblatt RB. Kinetics of T-lymphocyte subsets in the eyes of Lewis rats with experimental autoimmune uveitis. Cell Immunol 1985;96:430-4.
28 De Kozak Y, Sainte-Landy J, Benveniste J, Faure JP. Evidence for immediate hypersensitivity phenomena in experimental autoimmune uveitis. Eur J Immunol 1981;11:612-7.
29 Murray PI, Hoekzema R, Luyendijk L, Konings S, Kijlstra A. Analysis of aqueous humor imunoglobulin G in uveitis by enzyme-linked immunosorbent assay, isoelectric focussing and immunoblotting. Invest Opthalmol Vis Sci 1990;31:2129-35.
30 Murray PI, Hoekzema R, Van Haren MAC, De Hon FD, Kijlstra A. Aqueous humor interleukin-6 levels in uveitis. Invest Ophthalmol Vis Sci 1990;31:917-20.
31 Deschenes J, Char DH, Kaleta S. Activated T lymphocytes in uveitis. Br J Ophthalmol 1988;72:83-7.
32 Zierhut M, Mihalcic D, Duijvestijn AM, Thiel HJ. High endothelial venules in the eye. Invest Ophthalmol Vis Sci 1992;33/4:1245.
33 Egwuagu CE, Mahdi RM, Nussenblatt RB, Gery I, Caspi RR. Evidence for selective accumulation of Vß8+ T lymphocytes in experimental autoimmune uveoretinitis induced by two different retinal antigens. J Immunol 1993;151: 1627-36.

34 Rao NA, Naida YM, Bell R, Lindsey JW, Pararajasegaram G, Sun Y, Steinman L. Usage of T cell receptor ß-chain variable gene is highly restricted at the site of inflammation in murine autoimmune uveitis. J Immunol 1993;150:5716-21.
35 Kawano Y, Sasamoto Y, Kotake S, Thurau SR, Wiggert B, Gery I. Trials of vaccination against experimental autoimmune uveoretinitis with a T-cell receptor peptide. Curr Eye Res 1991;10:789-95.
36 Vandenbark AA, Hashim G, Offner H. Immunization with a synthetic T-cell receptor V-region peptide protects against experimental autoimmune encephalomyelitis. Nature 1989;341:541-4.
37 Smith JA, Whitcup S, Mahdi RM, Nussenblatt RB, Egwuagu CE. T cell receptor γ/δ usage in human ocular sarcoidosis. Invest Ophthalmol Vis Sci 1995;36/4:S537.
38 Whitcup SM, Egwuagu CE. The T-cell receptor in uveitis: molecular analysis of uveitogenic T-cells and the role of costimulatory signals in T-cell activation. In: Zierhut M, Saal JG, Thiel HJ, editors. Immunology of the joint and the eye. Buren, The Netherlands: Aeolus Press, 1996:205-20.
39 De Smet MD, Yamamoto JH, Mochizuki M, Gery I, Singh VK, Shinohara T, Wiggert B, Chader GJ, Nussenblatt RB. Cellular immune response of patients with uveitis to retinal antigens and their fragments. Am J Ophthalmol 1990;110:135-42.
40 Yamamoto JH, Minami M, Inaba G, Masuda K, Mochizuki M. Cellular autoimmunity to retinal specific antigens in patients with Behçet´s disease. Br J Ophthalmol 1993;77:584-9.
41 Chan CC, Palestine AG, Nussenblatt RB, Roberge FG, Ben Ezra D. Antiretinal auto-antibodies in Vogt-Koyanagi-Syndrome, Behçet´s disease, and sympathetic ophthalmia. Ophthalmology 1985;92:1025-8.
42 Gregerson DS, Abrahams IW, Thirkill CE. Serum antibody levels of uveitis patients to bovine retinal antigens. Invest Ophthalmol Vis Sci 1981;21:669-80.
43 La Hay E, Broersma L, Van der Gaag R, Baarsma GS, Rothova A, Kijlstra A. Does autoimmunity to S-antigen play a role in Fuchs´ heterochromic cyclitis? Br J Ophthalmol 1993;77:436-9.
44 Doeckes G, Van der Gaag R, Van Kooyk Y, Broersma L, Zaal MJM, Dijkman G, Fortuin ME, Baarsma GS, Kijlstra A. Humoral and cellular immune responsiveness to human S-antigen in uveitis. Curr Eye Res 1987;6:909-19.
45 Forrester JV, Stott D, Hercus K. Naturally occurring antibodies to bovine and human retinal S-antigen: a comparison between uveitis patients and healthy volunteers. Br J Ophthalmol 1989;73:155-9.
46 Uusitalo RJ, Stjernschantz J, Mahlberg K, Gregerson DS, Uusitalo H, Tallberg T, Palkama A. Serum antibody levels to S-antigen in children with chronic uveitis. Br J Ophthalmol 1985;69:212-6.
47 Murray P. Serum autoantibodies and uveitis. Br J Ophthalmol 1986;70:266-8.
48 Kijlstra A, Linssen A, Ockhuizen T, Baarsma GS, Dandrieu R, Tjoa T, De Graauw J, De Christiaans TM, Dekker-Saeys AJ, Feltkamp EW. Diagnostic tests in uveitis. In: Saari KM, editor. Uveitis update. Amsterdam: Elsevier Science Publ ,1984:421-4.
49 Rahi AHS, Holborow EJ, Perkins ES, Gungen YY, Dinning WJ. Immunologi-

cal investigations in uveitis. Trans Ophth Soc UK 1976;96:113-22.
50 Zierhut M, Klein R, Berg P, Waetjen R, Frohn A, Thiel JH, Pleyer U. Augenspezifische Autoantikörper bei Uveitis. Fortschr Ophthalmol 1989;86:482-5.

MICHAEL B. RAIZMAN

Immunopathology of ocular cicatricial pemphigoid

The purpose of this article is to review the immunopathology of ocular cicatricial pemphigoid. I will concentrate on a review of the literature describing the immunopathologic studies performed to date. However, since most readers may never have seen the clinical manifestations of this condition, and some may even have little knowledge of the eye, I will briefly describe the clinical aspects of the disease.

Ocular cicatricial pemphigoid is a rare disease that primarily affects the eyes, but also the skin and mucous membranes.[1] It is sometimes termed benign cicatricial pemphigoid, an unfortunate name, given the condition's ability to cause blindness and death. The clinical manifestations are often confused with other more common ocular conditions. Consequently, ocular cicatricial pemphigoid is probably under-diagnosed and more common than currently recognized.

The ratio of women to men with ocular cicatricial pemphigoid is about 1.6 to 1.8. Most individuals are around 65 to 70 when the disease is diagnosed. Cases have been found in patients as young as 25 and I have seen several patients who were first diagnosed with ocular cicatricial pemphigoid in their 90's. Depending on the series that you read, the conjunctival tissues of the eye are involved 66-100% of the time, the skin 9-24%, and the oral mucosa 50-91%. Naturally, the series reported by ophthalmologists have higher percentages of ocular involvement, while the dermatologists report more non-ocular involvement. In the eye, the primary findings are conjunctival hyperemia (see Fig. 1), scar formation with eventual adhesions (see Fig. 2), and secondary changes of dry eye, trichiasis (eye lashes turning in and rubbing the eye), and corneal scarring (see Fig. 3). The latter results in blindness in many cases. Ocular cicatricial pemphigoid can have periods of relapse and remission. Treatment

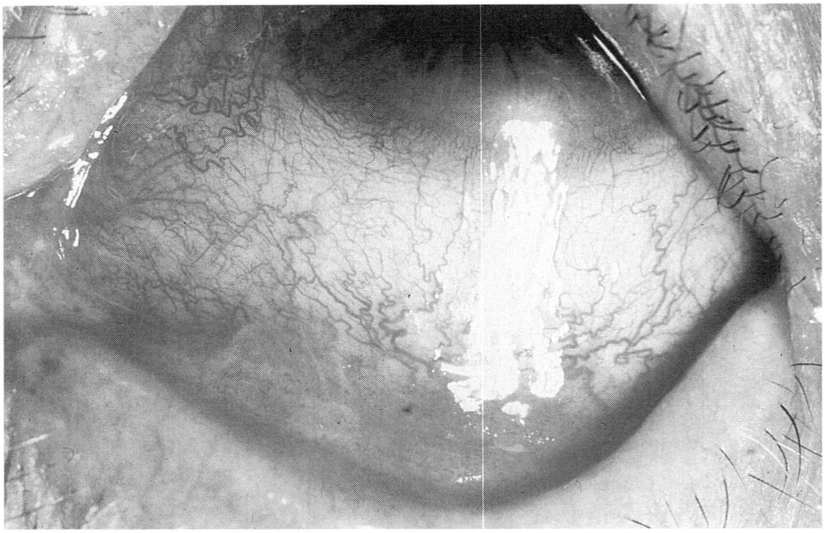

FIGURE 1 Conjunctival hyperemia on the ocular surface and lower eyelid conjunctiva in active ocular cicatricial pemphigoid. Note also the early white scars on the lower conjunctiva.

of the condition with systemic immunosuppressive therapy is usually successful. Some patients require lifelong therapy. The most

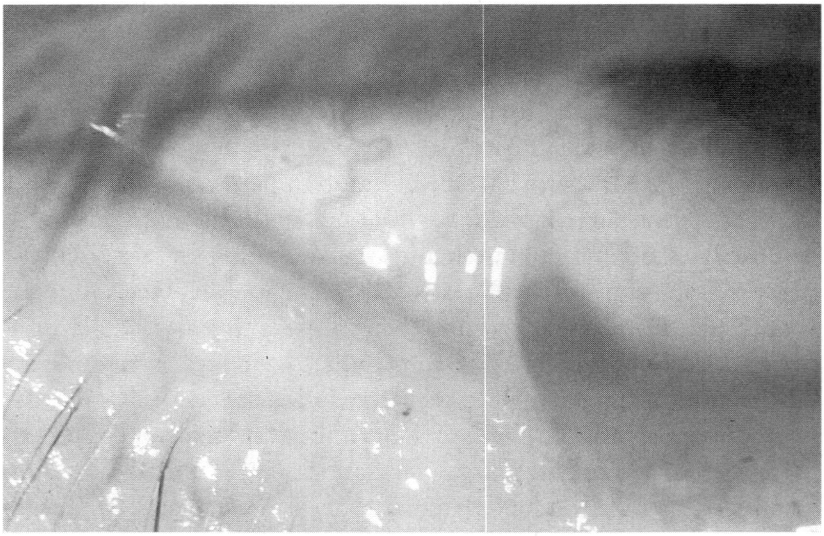

FIGURE 2 Symblepharon, or adhesions, of conjunctiva. In this case the adhesions have created an attachment from the eyelid margin to the surface of the eye.

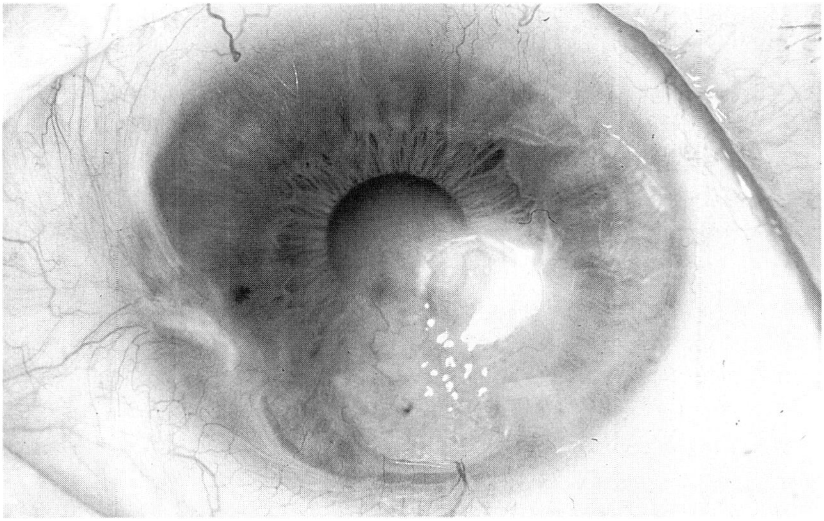

FIGURE 3 Corneal scarring after years of active ocular cicatricial pemphigoid. The eye no longer sees well because of irreversible scarring.

commonly used drugs include Dapsone and cyclophosphamide.

Studies on the immunopathology of ocular cicatricial pemphigoid are few and most that exist are limited in scope. Several reasons account for this. Ocular cicatricial pemphigoid is not known to occur in animals and there is no known animal model of the disease. The disease in humans is rare, limiting the number of patients available for study and biopsy. Only several centers in the world have large numbers of affected patients. The area of ocular involvement is small and the amount of tissue available for biopsy is small. In the eye, only the conjunctiva is affected, not the internal structures of the eye. Taking too large a biopsy of the conjunctiva would injure an already compromised ocular surface. Many biopsies are obtained after patients have received some therapy for the disease. Even though the therapy may not have been appropriate, it may have influenced the immunopathologic findings.

The histopathology of ocular cicatricial pemphigoid is characterized by a separation within the lamina lucida between the plasma membrane of basal epithelium and lamina densa. In the epithelium there is squamous metaplasia and keratinization. In the eye there is loss of conjunctival goblet cells. Acute and chronic inflammatory cells are common with polymorphonuclear leukocytes, lymphocytes, plasma cells, and macrophages.[2] In most cases, the disease has been present for years before biopsy, so chronic fibrosis will be found within the conjunctiva.

Mast cells are found in abundance in the conjunctiva of eyes with ocular cicatricial pemphigoid.[3] Many of these mast cells may be of the connective tissue type, though more precise identification remains to be performed. The significance of mast cells in this condition is unknown. Mast cells are known to proliferate in the presence of fibroblasts and their abundance in ocular cicatricial pemphigoid may be an epiphenomenon unrelated to disease pathogenesis. Mast cells can also induce fibroblast proliferation and in some conditions, such as neurofibromatosis, compounds that stabilize mast cells can reduce tumor size. Mast cell stabilizers are not known to help patients with ocular cicatricial pemphigoid, but adequate trials have not been performed.

The immunopathologic hallmark of ocular cicatricial pemphigoid is the binding of immunoglobulins and complement components to the epithelial basement membrane of skin, oral mucosa, and conjunctiva.[1,2] IgG, IgA, and C4 are most commonly found. The antigens of ocular cicatricial pemphigoid and bullous pemphigoid are distinct but both are localized within the lamina lucida of the dermal-epidermal junction. The bullous pemphigoid antigen is located closer to the basal epithelial cells while the ocular cicatricial pemphigoid antigen is closer to the lamina densa. Occasionally, antibody may bind to the epithelium of tissue from patients with ocular cicatricial pemphigoid. Circulating antibodies to the epithelial basement membrane zone can be found in 50% of patients. There is an association with ocular cicatricial pemphigoid and HLA-DQw3.[4] This was carried with linkage to DR4 in 19 out of 20 patients tested.

Foster and his associates have performed and reported some of the most important work describing the conjunctival immunopathology of ocular cicatricial pemphigoid. In a study by Rice and Foster,[2] published in 1994, they compared conjunctiva from eyes with active ocular cicatricial pemphigoid versus normal controls. In addition to the expected finding of immunoglobulin and complement deposition at the basement membrane zone of the conjunctival epithelium in the ocular cicatricial pemphigoid samples, they described inflammatory cells in the conjunctival epithelium and substantia propria. In the epithelium they found significantly more T helper cells (CD4+), dendritic cells (CD1+), macrophages (CD14+, Mac-1+), and a higher helper/suppressor (CD 4+/8+) ratio than in controls.

In the substantia propria of ocular cicatricial pemphigoid samples, a large increase in infiltrating T cells (CD 3+, CD5+), T helper cells (CD 4+), and T suppressor cells (CD 8+) was found. A smaller increase in the number of macrophages (CD14+, Mac-1+) and dendritic cells (CD1+, HLA-DR+) was noted. Ten percent of cells from patients with ocular cicatricial pemphigoid expressed the IL-2 receptor protein (CD25+), indicating T cell activation. This work confirmed and expanded upon the work of Sacks et al.[5], suggesting a role for T cells in ocular cicatricial pemphigoid.

In 1993, Bernauer et al.[6] published data comparing active, subacute, and chronic ocular cicatricial pemphigoid. In the active disease there were abundant macrophages and neutrophils. All groups had increased T cells but the highest number was found in the group with subacute disease. In the groups with subacute and chronic ocular cicatricial pemphigoid, there were more CD8+ than CD4+ lymphocytes. In the group with active disease the numbers of CD8+ and CD4+ lymphocytes were equal. The IL-2 receptor was found on 5% of cells. There was increased MHC class II expression in all groups. B cells and NK cells were not increased. Interestingly, TGF-ß staining was noted in active disease. Ocular cicatricial pemphigoid is a cicatrizing disease and TGF-ß may be fibrogenic. Lee et al.[7] reported that serum levels of TNF-alpha and IL-6 were elevated in 35 patients with ocular cicatricial pemphigoid, suggesting that ocular cicatricial pemphigoid is a systemic disease, and supporting a role for cytokines in the pathogenesis.

Ahmed et al.[8] have searched for the ocular cicatricial pemphigoid antigen or antigens and reported, in 1993, their findings utilizing human sera and COLO and SCaBER tumor cell lines. They employed a sensitive radioimmunoblot assay. Sera from patients with ocular cicatricial pemphigoid recognized peptide(s) present in 230, 205 and 160 kD proteins in lysates from tumor cells. They also demonstrated that the ocular cicatricial pemphigoid antigen(s) appear(s) to be distinct from the bullous pemphigoid antigen(s).

Smith et al.[9] looked at patients with ocular cicatricial pemphigoid with pure ocular involvement. All of the patients had IgG and/or IgA at the conjunctival basement membrane zone. The patients also had elevated IgA levels in the serum. Immunoblots detected IgA binding to a 45-kD basement membrane zone antigen. This antigen was distinct from antigens for other cutaneous diseases.

An interesting study reported by Chan et al.[10] looked at indirect immunofluorescence on a salt-split-skin substrate to detect circulating antibasement antibodies, direct immunofluorescence, and serologic reactivity against cultured keratinocyte antigens in patients with ocular cicatricial pemphigoid and bullous pemphigoid. They determined that a subset of patients with only ocular lesions of cicatricial pemphigoid exhibited significantly lower *in vivo* deposits of IgG and C3, higher deposits of fibrin, virtual absence of circulating antibodies, and negative serologic reactivity to bullous pemphigoid antigens. The group of cicatricial pemphigoid patients with combined oral and cutaneous lesions were more similar to patients with bullous pemphigoid. These findings suggest an alternative classification of these cicatrizing disorders.

Mention must be made of a form of ocular cicatricial pemphigoid that arises after the use of oral medications or topical eye drops.[1] Various eye drops (especially anti-glaucoma preparations) and oral practolol (a beta-blocker) have been known to induce ocular changes clinically indistinguishable from ocular cicatricial pemphigoid. So-called drug-induced pemphigoid does not always improve with cessation of the implicated medication. The disease is usually responsive to systemic immunosuppressive therapy. These patients also have histopathologic, electron microscopic, and immunopathologic changes in the conjunctiva that are identical to those found in ocular cicatricial pemphigoid. The disease can affect one or both eyes (even if medications are used only in one eye). Most cases have no extra-ocular signs. Explanations include a possible coincidental appearance of ocular cicatricial pemphigoid in individuals using these drugs, implying no cause and effect. It is possible that the drug accelerates or unmasks a predisposition to ocular cicatricial pemphigoid in susceptible individuals. The drug may actually cause ocular cicatricial pemphigoid or cause scarring that leads to all the pathologic changes seen in ocular cicatricial pemphigoid. It is possible that ocular cicatricial pemphigoid may cause problems like glaucoma that lead to the need for drug therapy, making the use of drugs more common in those with ocular cicatricial pemphigoid.

How can we combine the limited information we have on the immunopathology of ocular cicatricial pemphigoid to propose a mechanism of the pathologic changes? It is possible that there is a primary autoimmune reaction with antibody to the epithelial basement membrane zone or conjunctival epithelial protein. This may be followed by complement activation. There is clearly migration of inflammatory cells in the affected tissue, perhaps in response to chemotactic factor release. IL-1 and other cytokine release by macrophages may be accompanied by recruitment of other inflammatory cells and class II induction. Damage to epithelium results with fibroblast proliferation and collagen secretion, resulting in the characteristic clinical finding of cicatrization. Our current understanding of the immunopathology of ocular cicatricial pemphigoid has not yet led to new methods of treatment or prevention of this potentially blinding and life-threatening disease.

References

1 Mondino BJ. Cicatricial pemphigoid and erythema multiforme. Ophthalmology 1990;97:939-52.

2 Rice BA, Foster CS. Immunopathology of cicatricial pemphigoid affecting the conjunctiva. Ophthalmology 1990;97:1476-83.

3 Hoang-Xuan T, Foster CS, Raizman MB, Greenwood B. Mast cells in conjunctiva affected by cicatricial pemphigoid. Ophthalmology 1989;96:1110-4.

4 Ahmed AR, Foster CS, Zaltas M, Awdeh Z, Alper CA, Yunis EJ Association of DQw7 (DQB1*0301) with ocular cicatricial pemphigoid. Proc Natl Acad Sci USA 1991;88:11579-82.

5 Sacks EH, Jakobiec FA, Wieczorek R, Donnenfeld E, Perry H. Knowles Jr DM. Immunophenotypic analysis of the inflammatory infiltrate in ocular cicatricial pemphigoid. Further evidence for a T cell-mediated disease. Ophthalmology 1989;96:236-43.

6 Bernauer W, Wright P, Dart JK, Leonard JN, Lightman S. The conjunctiva in acute and chronic mucous membrane pemphigoid. An immunohistochemical analysis. Ophthalmology 1993;100:339-46.

7 Lee SJ, Li Z, Sherman B, Foster CS. Serum levels of tumor necrosis factor-alpha and interleukin-6 in ocular cicatricial pemphigoid. Invest Ophthalmol Vis Sci 1993;34:3522-5.

8 Mohimen A, Neumann, Foster CS, Ahmed AR. Detection and partial characterization of ocular cicatricial pemphigoid antigens on COLO and SCaBER tumor cell lines. Curr Eye Res 1993;12:741-52.

9 Smith EP, Taylor TB, Meyer LJ, Zone JJ. Identification of a basement membrane zone antigen reactive with circulating IgA antibody in ocular cicatricial pemphigoid. J Invest Dermatol 1993;101:619-23.

10 Chan LS, Yancey KB, Hammerberg C, Soong HK, Regezi JA, Johnson K, Cooper KD. Immune-mediated subepithelial blistering diseases of mucous membranes. Pure ocular cicatricial pemphigoid is a unique clinical and immunopathological entity distinct from bullous pemphigoid and other subsets identified by antigenic specificity of autoantibodies. Arch Dermatol 1993;129:448-55.

ARTHUR MELMS, ROBERT WEISSERT, ALEXEJ SCHMIDT, WOLFGANG WIENHOLD, GEORG MALCHEREK, CLAUDIA MÜLLER AND GÜNTHER JUNG

Myasthenia gravis as a model autoimmune disease: T cell epitopes of the acetylcholine receptor

Introduction: Cinical and pathophysiological aspects

Myasthenia gravis (MG) is an acquired autoimmune disorder of neuromuscular transmission. The clinical symptoms of MG are abnormal weakness and fatigability of muscles, often present in a characteristic distribution. A limited disease restricted to extraocular muscles, ocular myasthenia, is characterized by ptosis and diplopia. Often the disorder becomes generalized and involves muscles in the face and neck, arms and legs. Exacerbation of generalized muscle weakness by infection, physical exhaustion or other factors may lead to acute myasthenic crisis which is a life-threatening situation and requires prompt intensive-care treatment of respiratory failure and imminent aspiration. The principles of diagnosis and current therapy of MG have recently been reviewed.[1] The elucidation of the target antigen, the acetylcholine receptor (AChR), and details concerning the pathological mechanisms have made MG the prototype for the study of human autoantibody-mediated diseases (see Table 1).

The acetylcholine receptor is the autoimmune target in MG

The AChR is the target of polyclonal IgG-antibodies which aggregate receptor complexes and accelerate their internalization (modulating antibodies), eventually reducing their half-life. Only a minority of AChR-antibodies directly interfere with agonist binding and

TABLE 1 Myasthenia gravis: a prototype autoimmune disease.

Autoimmune target	acetylcholine receptor (AChR) of skeletal muscle
Pathophysiology and mechanisms	loss of AChR by modulating auto-antibodies; impairment of neuromuscular transmission
Immunopathology	deposits of immunoglobulins and terminal complement components; focal destruction of muscle end plate architecture (membrane attack complex)
Animal model	immunization with AChR induces experimental autoimmune myasthenia gravis (EAMG) in mice, rats, rabbits, monkeys.
Transfer of myasthenic symptoms	by IgG from MG patients or from animals with EAMG or monoclonal antibodies to AChR.
Immunotherapy	plasma exchange is temporarily effective; immunotherapy is effective.

block neuromuscular transmission. In addition, the end-plate architecture is focally destroyed by complement activation, and as a consequence, severly damaged myasthenic end-plates have lost about 70-90% of their AChR. In physiological terms, there is a critical reduction of the *safety factor of neuromuscular transmission* and neuromuscular transmission fails under repetitive stimulation.

The AChR is a ligand-gated ion channel formed by five polypeptide subunits $\alpha_2\beta\gamma\delta$ in a barrel-like arrangement.[2] The AChR subunits are coded by different gene loci but extensive homology suggests a common ancestral gene. There is also sequence homology across species, e.g. there is about 80% sequence identity between the α-subunits from the electric organ of fish and human muscle. All subunits are immunogenic and antibodies cross-reacting between species induce experimental autoimmune myasthenia gravis (EAMG) in animals[3] which have clinical signs, end-plate pathology and pharmacological response to cholinesterase inhibitors like patients with MG.

Experimental autommune myasthenia gravis (EAMG)

EAMG has been a useful model to explore experimental immunotherapies. Strategies to deplete autoantibodies and eliminate antibody-producing B-cells but also T cell directed therapies have been effective (see Table 2). The depletion of T lymphocytes prevents the production of autoantibodies to AChR.[4] Similarly, inhibition of antigen recognition of T helper cells by monoclonal antibodies to MHC class II molecules suppresses the immune response to AChR and prevents EAMG *in vivo*.[5] Other strategies directed at the T cell receptor of AChR-specific T cells have also been shown to interfere with T cell activation and the development of EAMG.[6]

The thymus in MG

Although the origin of the autoimmune response to AChR remains elusive, it has long been recognized that most MG patients have thymus abnormalities, including lymphoepithelial tumors of the thymus. MG is the most frequent paraneoplastic syndrome in thymoma patients. Several pieces of evidence support the pivotal role of the thymus and thymus-dependent immune responses in the pathogenesis of MG.[7] Thymus and thymoma in MG are enriched with T cells recognizing AChR.[8,9] As a hallmark, in the non-neoplastic MG thymus, germinal centers arise and intrathymic B lymphocytes which produce substantial amounts of AChR-autoantibodies show somatic hypermutation, a process which requires T cell help. AChR is expressed on thymic myoid epithelial cells, and thymic myoid cells are found in intimate contact with interdigitating reticulum cells (antigen-presenting cells) in a thymic compartment predominated by mature CD3-positive T cells.[8] However, there is no evidence that myoid cells themselves are targets of an autoim-

TABLE 2 Strategies in the Treatment of Experimental Autoimmune Myasthenia gravis.

Strategy	Reference
• anti-MHC-class-II-antibodies	Waldor et al., 1983
• immunotoxins (AChR-ricin-conjugates)	Killen et al., 1985
• anti-idiotypic-antibodies	Souroujon et al., 1986
• AChR-design-peptides or anti-T-cell-receptor antibodies block antigen-presentation	Brocke et al., 1990
• AchR-peptide-conjugates	Atassi et al., 1992
• AChR fixed to specific B-lymphocytes	Reim et al., 1992
• soluble AChR-peptide MHC-complexes	Nicolle et al., 1993
• oral tolerance with Torpedo-AChR	Wang et al., 1993

mune response in the thymus.[10]

Thymoma cells, i.e. neoplastic thymic epithelial cells, do not express AChR polypeptides but they share determinants with AChR[11] and are able to present antigens, including the AChR, and stimulate T cells.[12] These observations in the context of a paraneoplastic immune response support the concept of immunological mimicry[13] between a thymoma antigen and the AChR. Taken together in the clinical setting, this argues in favor of thymectomy as a therapeutic measure as it is often followed by significant reduction in AChR autoantibody titer and amelioration of the course of MG.

AChR-specific T helper cells

T helper cells are required for sustained antibody production and secretion of high affinity IgG antibodies by B lymphocytes. During the past decade, MG has received much attention since it offered a unique opportunity to analyze T and B cell responses to a well defined autoantigen. By elucidating the role of AChR-specific T cells, it was hoped to design selective immunotherapies[14] to control the autoimmune deviation by tuning down autoreactive AChR-specific T helper cells.[6] CD4-positive T cell lines were isolated from the peripheral blood and the thymus of MG patients using torpedo-AChR as a substitute for human muscle receptor.[8,9,15] Such T cell lines were useful tools to study antigen recognition in defined cell populations (see Figs. 1A and B). They were shown to provide 'helper signals' for the production of AChR-antibodies in vitro[16] and had a cytokine profile of the Th1 phenotype.[17] Although there was some cross-reactivity between AChR from torpedo and human muscle, the majority of AChR-specific T cells exclusively recognized the selecting antigen (human or torpedo AChR, see Fig. 2). Similarly, only a small fraction of AChR-antibodies from MG-patients recognized AChR from torpedo, so there was a need to use the authentic human antigen for further studies. Progress in bacterial expression systems made available recombinant mammalian and human AChR subunits which allowed one to map regions of prominent T cell epitopes.[18-20] Remarkably, AChR-specific T cells were not only found in MG patients but were also present in normal healthy volunteers, indicating potentially autoreactive T cells in the peripheral T cell repertoire of normal individuals.[20-22]

T cells recognize antigen fragments generated by intracellular

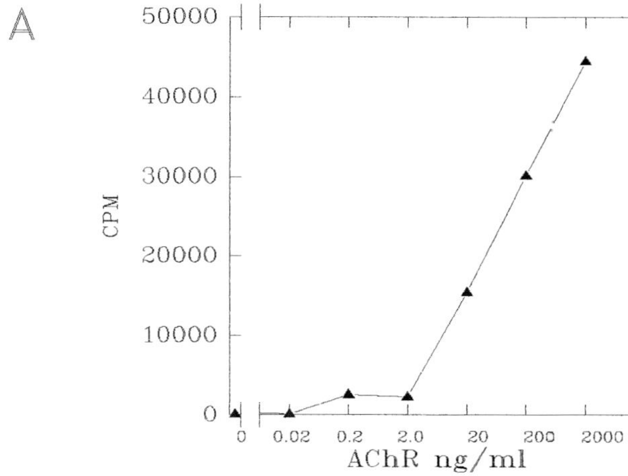

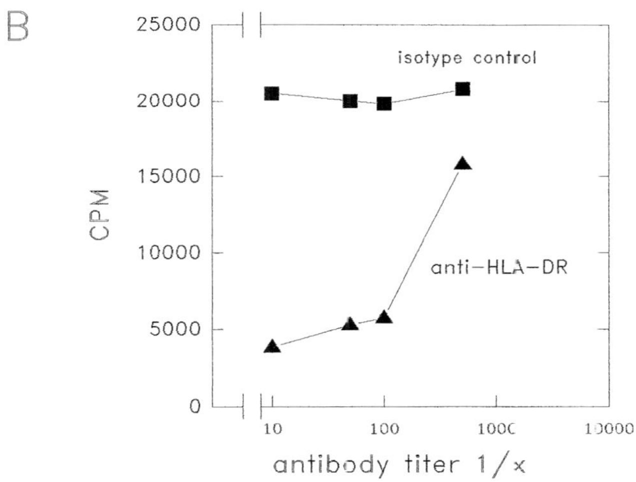

FIGURE 1 A: Dose-response of a CD4-positive T cell clone to purified AChR presented by autologous antigen-presenting cells. 1pM AChR is about 0.28 ng/ml protein.
B: CD4-positive T cells recognize AChR in the context of HLA-DR molecules. Preincubation of antigen-presenting cells with a monoclonal antibody directed at HLA-DR determinants interferes with AChR-induced T cell proliferation. The presence of an irrelevant antibody as isotype control has no effect.

degradation in antigen-presenting cells;[23] hence, they may respond to determinants from all over the molecule, including epitopes on intra-/extra- and transmembrane segments of the AChR. In an am-

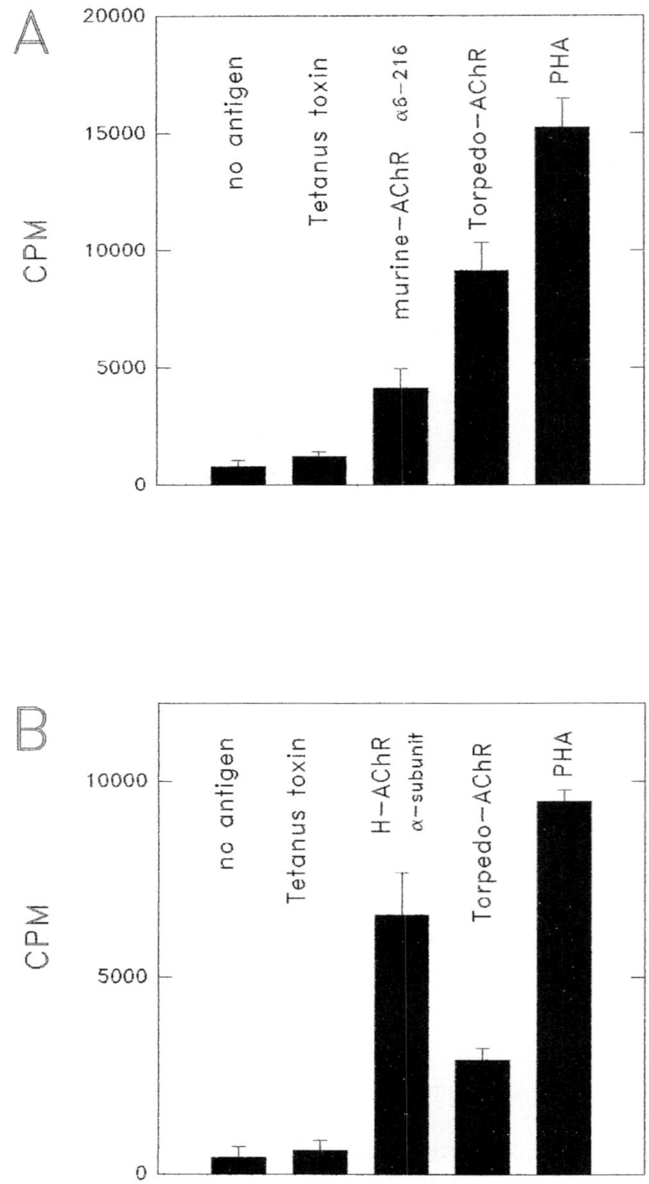

FIGURE 2 Limited cross-reactivity of T cells selected with Torpedo-AChR (A) or recombinant human AChR α-subunit (B). In majority, T cells were biased to respond to the selecting antigen and showed little cross-reactivity to AChR from other sources.

bitious approach to identify T cell epitopes, a battery of overlapping peptides was synthesized, spanning all subunits of the human AChR, which revealed a striking heterogeneity of AChR peptides that evoked T cell responses.[24] Although AChR is a highly immunogenic protein antigen, this is in some contrast to a more restricted T cell response reported from exogenous antigens. It has been argued that the use of predefined peptide pools can easily lead to an overestimate of the heterogeneity of epitopes and many peptides may not be generated by proteolytic degradation of the original antigen. This issue requires further investigation into the principles of antigen processing and presentation[25] and the peptides actually presented by MHC class II restriction molecules.

Immunogenetic aspects of disease susceptibility

MG, like many autoimmune diseases, is associated with certain HLA haplotyes like HLA-B8 and HLA-DR3, but this differs between various ethnic groups as well as among patients with different clinical presentations (see Table 3).[26] Patients with paraneoplastic MG show no consistently positive HLA-association but it was observed that the presence of antibodies to striated muscle antigens other than AChR (as a marker for a thymoma) was inversely correlated with the predisposing haplotype A1/B8/DR3.[27] The association of MG in young adult caucasians with HLA-DR3 prompted us to examine the role of HLA-DR3 as an antigen-presenting molecule for AChR-specific T cells. There is no evidence that DR3 (the new nomenclature is DR17 or DRB1*0301) predominates the restriction of AChR-specific T cells *in vitro*, but this might help to explain whether there is also a functional contribution of DR3(17) to the pathogenesis of MG.

TABLE 3 Heterogeneity of Myasthenia gravis (Compston et al. 1980).

clinical myasthenia	*age at onset*	*thymus pathology*	*AChR-ab. titer*	*HLA- association*
generalized	<40 years	hyperplasia	high (>6 nM)	A1, B8, DR3
generalized	>40 years	involution	low (<6 nM)	A2, B7, DR2
generalized	any age	thymoma	high	none
ocular	any age	not avail.	low or neg.	not examined
generalized	any age	involution	negative (<0.4)	not examined

MHC class II molecules are peptide receptors which display, on the cell surface, peptide fragments which can be recognized by T helper cells.[28,29] The binding site (groove) of MHC class II molecules is polymorphic and consequently, peptide ligands have to follow allele-specific constraints to succeed in binding. In essence, the peptide must contain compatible amino acids of a sequence motif which is a pattern of 3 to 4 conserved residues of a nine-mer core segment fitting the binding groove.[30]

In laboratory animals, it has been shown that 'immune response genes' now identified as MHC class II molecules control susceptibility and resistance to experimental autoimmune myasthenia gravis (EAMG).[31] The bm-12 mutation of the murine MHC class II molecule I-A has altered the peptide binding site; hence, animals with the bm-12 mutation respond to a different set of AChR-peptides compared to the wild type and mutant animals are resistant to EAMG.[32] This emphasises the strong influence of MHC class II molecules on determinant selection, regulation of the immune response, and susceptibility to autoimmune diseases.

We adopted the approach originally developed for the study of MHC class I associated peptides[33] and analyzed pools of peptides released from purified HLA-DR3(17) molecules by micro-Edman degradation and amino acid analysis. We extracted a unique motif for DR3(17)-ligands that was confirmed by a number of prominent natural ligands.[34] Specific contact sites were systematically evaluated by alanine-substituted analogues of natural ligands[35] which confirmed and extended a similar motif reported earlier.[36] In essence, the DR17-motif has four MHC contact sites arranged in a pseudo-symmetric pattern at positions 1-4-6-9. There is an allele-specific anchor at position 4 which is almost exclusively aspartate (D) and which is perfectly suited to interact with a complementary pocket on the floor of the DR17 binding groove (see Table 4).[34] This detailed information provided the basis to select candidate sequences biased for binding to HLA-DR3(17) and eventually forming T cell epitopes. Meanwhile, ligand motifs have been determined for many MHC-alleles. A motif and ligand register would be a useful guide for a straightforward selection of T cell determinants from protein sequences.[30]

DR3(17)-binding motif (peptide core)	1 * * 4 * 6 * * 9	C$_{50\%}$ inhibition of DR17-agonist- binding (μM)
	L D K Y	
	I R F	
	F E R	
	M Q hydrophobic	
	V aliphatic/ aromatic residues	
natural DR17-ligand (agonist)		
ApoB (2877-92)	I S N Q L T L D S N I K Y F H K	**1.0**
acetylcholine receptor peptides (motif peptides)		
α7-24	L V A K L F K D Y S S V V R P V E D	**38**
α37-54	L I Q L I N V D E V N Q I V T T N V	>200
α63-80	Y N L K W N P D D Y G G V K K I H I	>200
α82-99	S E K I W R P D L V L Y N N A D G D	**46**
α131-148	I V T H F P F D E Q N C S M K L G T	>200
α145-162	K L G T W T Y D G S V V A I N P E S	**40**
α231-248	L V F Y L P T D S G E K M T L S I S	>200
α310-327 human	N W V R K V F I D T I P N I M F F S	**2**
α310-327 torpedo	Q W V R K I F I D T I P N V M F F S	**2**
α338-355	D K K I F T E D I D I S D I S G K P	**70**
α382-399	I A E T M K S D Q E S N A A A E W K	**127**
α400-417	K Y V A M V M D H I L L G V F M L V	**4**

TABLE 4 Binding capacity of DR-17-motif peptides from human acetylcholine receptor α-subunit.
The nine-mer core of the DR 17-ligand motif and the presence of anchor motif residues in AChR-peptides are indicated in bold. The binding capacities of motif peptides to purified HLA-DR3(17) molecules were measured in competition to a biotinylated high affinity natural ligand, ApoB(2877-92) as described. Results (in μM) were calculated for 50% inhibition of agonist binding as described (35). Binding capacity of the agonist peptide was determined against an unbiotinylated analogue.
(Single letter symbols for amino acids: alanine A; arginine R; asparagine N; aspartic acid D; cysteine C; glutamine Q; glutamic acid E; glycine G; histidine H; isoleucine I; leucine L; Lysine K; methionine M; phenylalanine F; proline P; serine S; threonine T; tryptophane W; tyrosine Y; valine V.

AChR-peptides binding to HLA-DR3(17)

The α-subunit of the AChR is the most immunogenic in humans and experimental animals.[37,38] Screening for DR17-motifs, there are eleven candidates along the molecule which, in a nine-mer core segment, have hydrophobic aliphatic or aromatic residues at relative position 1 and the DR17-specific anchor, aspartate (D), at relative position 4. Some peptides also have compatible residues at positions 6 and/or 9 fitting the DR17-binding groove. The binding of these AChR motif-peptides to HLA-DR17 was analyzed at endosomal pH 5.0 in competition with a high affinity natural ligand, apoB(2877-92), as described in Table 4.[35] The results show that motif screening is a powerful tool to predict peptides with specific MHC-binding capacity. The majority of motif-peptides had good DR17-binding capacities, some of them were as active as natural high affinity ligands.

MHC-binding capacity is certainly a major prerequisite for peptides to succeed in binding. T cells, however, are tolerant and ignore high affinity peptides from endogenous proteins like peptides from apolipoprotein, transferin receptor and other natural DR17-ligands. Therefore, selection and shaping of the T cell repertoire is essential for the physiology and function of the adaptive cellular immune system.

AChR motif-peptide forming T cell epitopes

In primary cultures stimulated with single peptides, we noted that virtually all motif-peptides evoked responses that exceeded the controls. A typical example is shown in Fig. 3. In general, the responses were rather low compared to stimulation with recall antigens. Although there was an individually different pattern of stimulating peptides, there was no difference between DR3-positive healthy individuals and DR3-positive MG patients. AChR-peptide responses were more frequent and stronger in normal DR3-positive donors than in MG patients (who were not on immunosupression). Motif-peptides α231-248, α310-327 and α338-355 were the prominent T cell determinants recognized in the DR17-positive population. By contrast, determinants of the extra-cellular part of the α-subunit (α1-216) evoked less frequent responses. Few peptides evoked responses in non-DR3(17) controls, which can be explained by the presence of additional motifs in these peptide sequences. For example, there is a cluster of motifs in α310-327 compatible with binding to DR1, DR3 and DR4 (see Table 5).

T lymphocytes in primary cultures are highly polyclonal and proliferation to a single peptide does not necessarily correlate with the response to deteminants generated by endogenous antigen processing. We addressed this issue using T cell clones selected with AChR

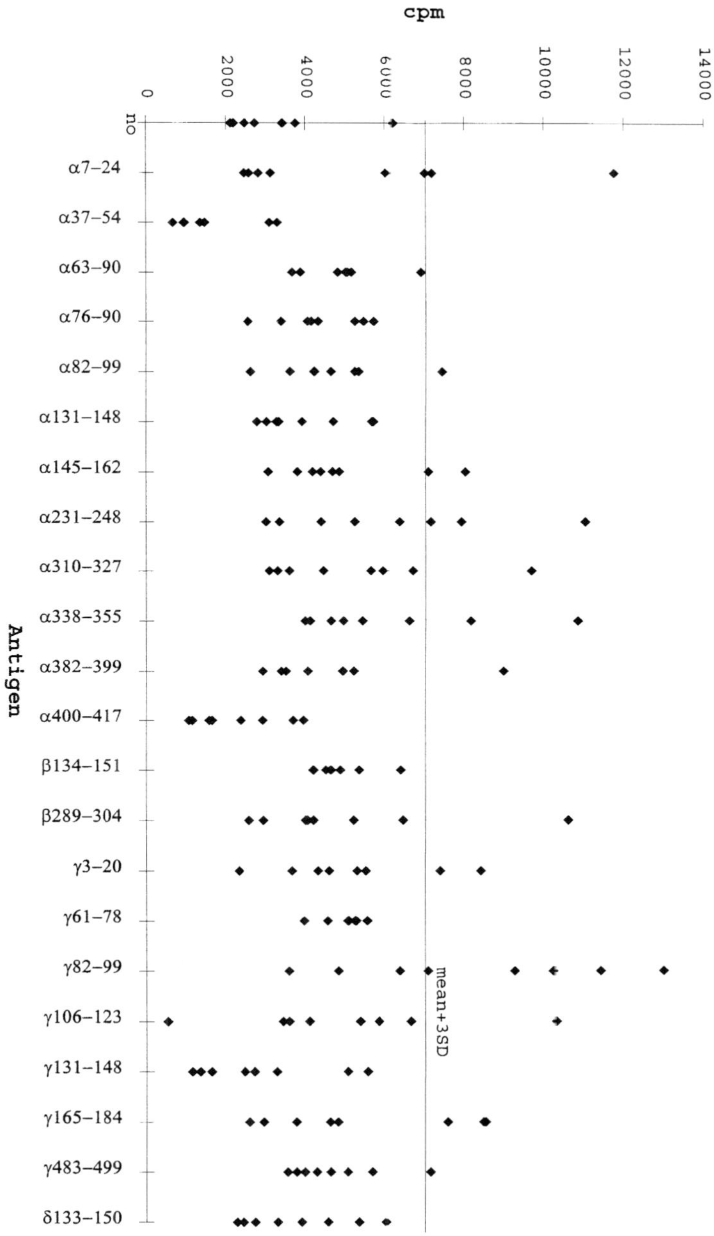

Fig. 3: DR17-motif peptides from human AChR induce T cell proliferation in primary cultures from a DR3(17)-positive donor. Eight replicate cultures, each containing a single peptide (10 μg/ml) and 100,000 PBMC, were cultured for 6 days and ^{3}H-thymidine incorporation was measured in a β-counter. Values exceeding the mean of the controls (eight replicates) plus 3SD were considered positive.

310 11 12 13 14 15 16 17 18 19 20 21 22 23 24 25 26 27		
N W V R K V F I D T I P N I M F F S	▼ ① ④ ⑥ ⑨ (V15, I17, T19, I23)	DR3(17); DR4
N W V R K V F I D T I P N I M F F S	▼ ① ④ ⑨ (V12, V15, I20)	DR1
N W V R K V F I D T I P N I M F F S	▼ ① ⑥ ⑨ (F16, P21, M24)	DR1

▼ indicates MHC-peptide contact site

TABLE 5 Cluster of motifs in AChR α310-327.

α231-248, α310-327, and α338-355 from DR17-positive high responders. Fig. 4 shows T cells responding vigorously to AChR peptide α338-355 while ignoring all the longer antigen fragments containing this particular sequence. Similarly, another T cell clone raised with AChR α310-327 responded exclusively to the peptide but not to recombinant antigens or torpedo-AChR. This peptide has previously been reported as a T cell epitope from several different laboratories[24,39,40] and happens to be one of the best DR17-binding peptides that we have found (see ref. 35 and Table 1). Apparently, these peptide-selected T cells do not reflect the true repertoire of AChR-specific T helper cells. As they do not recognize the antigen proper, they are unlikely to be involved in the pathogenesis of MG. Supporting the view of such AChR-peptides as non-physiological processing products, it was reported that digestion of recombinant polypeptides by trypsin, which is not involved in the physiological antigen degradation in lysosomes, created products that were active and stimulated α310-327-specific T cells *in vitro*.[40]

This situation is in marked contrast to our results with DR17-motif peptides from a foreign antigen, tetanus toxin: choosing a set of eleven DR17-motif-peptides, there was a perfect correlation between the binding performance of tetanus toxin motif peptides and their nature as T cell determinants. In particular, on the clonal level, motif-peptides could be shown to be equivalents of naturally-processed fragments of tetanus toxin (Malcherek et al., manuscript submitted).

Currently, the physiological role of T cells recognizing AChR

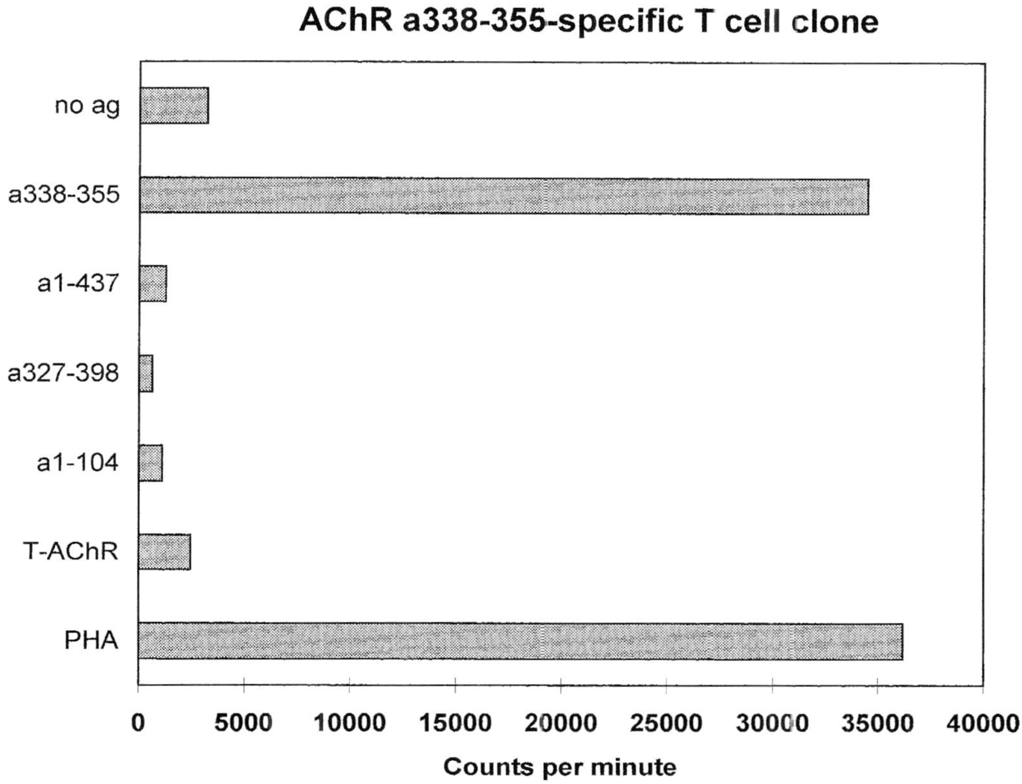

FIGURE 4 T cells recognizing AChR α338-355 motif peptide do not recognize naturally processed AChR fragments containing this epitope. Cloned T cells specific for AChR α338-355 motif peptide were challenged with recombinant polypeptides of the human AChR α-subunit or AChR from torpedo (T-AChR) but failed to respond to fragments containing this sequence. Bars indicate T cell proliferation measured as counts per minute of ^{3}H-thymidine incorporation.

motif-peptides described above remains unclear. It is reasonable to assume that not all possible self-antigen determinants are generated, so T cell tolerance will be restricted to those epitopes that were presented physiologically.[41] Sercarz et al.[42] have reported antigenic spreading to cryptic epitopes in autommunity and this concept requires further investigation.[43] Motif-peptides like the ones described here are biased to have good MHC binding capacities. Recent observations suggest that autoreactive T cells may have escaped the thymic selection process by recognizing low affinity ligands.[44,45] Hence, the AChR-motif peptides described here may represent some of the flexibility or degeneracy of an encyclopedic T cell repertoire and there is evidence from protein data bases that motif peptides and compatible analogues of autoantigens may be present in numerous human pathogens. Progress in the molecular resolution of antigen recognition will further challenge the current view about autoimmunity and protective immune responses.

Conclusions

We have applied the DR17-ligand motif to select, from the sequence of the AChR α-subunit, a set of peptides of which the majority had good binding capacities. Hence, the DR17-motif in peptide sequences correlated with DR17-binding. All AChR-motif peptides evoked significant proliferation in peripheral blood lymphocytes from a panel of DR17-positive donors. Whereas high affinity motif peptides from tetanus toxin were found to be equivalent to naturally processed antigen fragments, T cells selected with AChR peptides did not recognize AChR polypeptides containing the respective sequences. There is evidence that some DR17-motif peptides of AChR are not generated by physiological processing of AChR polypeptides in antigen-presenting cells. Therefore, peptide-specific T cells do not reflect the true repertoire of autoreactive AChR-specific T helper cells in myasthenia gravis. With regard to peptides forming T cell determinants, it appears that high affinity peptides of AChR do not follow the rules of foreign antigenic peptides. So far, this is in accordance with the view that autoreactive T cells may have escaped thymic tolerance induction and may recognize low affinity determinants. Given these findings, it would be appropriate to re-examine the role of T lymphocyes in the pathogenesis of MG.

Acknowledgements

We appreciate reagents, help and comments from Ralph Schoepfer, Jon Lindstrom, Kirsten Falk, Olaf Rötzschke, Hans-Georg Rammensee, Hubert Kalbacher, Volker Gnau and Stephan Stevanovic. We thank U. Gern for expert technical assistance. Work in the laboratory of A. M. was supported by the DFG, Sonderforschungsbereich 120 and a grant from the Sandoz Stiftung für Therapeutische Forschung. G. M. was supported by a fellowship from the fortüne-Programm of the University of Tübingen.

References

1 Drachman DB. Myasthenia gravis. N Engl J Med 1994;330:1797-1810.

2 Changeux JP, Devillers-Thiery A, Chemouilli P. Acetylcholine receptor: an allosteric protein. Science 1984;225:1335-45.
3 Patrick J, Lindstrom J. Autoimmune response to acetylcholine receptor. Science 1973;180:871-2.
4 Lennon VA, Lindstrom JM, Seybold ME. Experimental autoimmune myasthenia gravis: Cellular and humoral immune responses. Ann NY Acad Sci 1976;274:283-99.
5 Waldor MK, Sriram S, McDevitt HO, Steinman LS. In-vivo therapy with monoclonal anti-I-A antibody suppresses immune response to acetylcholine receptor. Proc Natl Acad Sci USA 1983;80:2713-7.
6 Brocke S, Dayan N, Steinman L, Mozes E. Inhibition of T cell proliferation specific for acetylcholine receptor epitopes related to myasthenia gravis with antibody to the T cell receptor or with competitive synthetic polymers. Int Immunol 1990;2:735-42.
7 Wekerle H. The thymus in myasthenia gravis. Ann NY Acad Sci 1993;681: 47-55.
8 Melms A, Schalke B, Kirchner T, Albert E, Müller-Hermelink HK, Wekerle H. Thymus in myasthenia gravis: Isolation of T lymphocyte lines specific for the nicotinic acetylcholine receptor from myasthenic patients. J Clin Invest 1988;81:902-8.
9 Sommer N, Willcox N, Harcourt GC, Newsom-Davis J. Myasthenic thymus and thymoma are selectively enriched in acetylcholine receptor-reactive T cells. Ann Neurol 1990;28:312-9.
10 Schluep M, Willcox N, Vincent A, Dhoot G, Newsom-Davis J. Acetylcholine receptors in human thymic myoid cells in situ: an immunohistological study. Ann Neurol 1987;22:212-22.
11 Marx A, O'Connor R, Geuder KI, Hoppe F, Schalke B, Tzartos S, Kalies I, Kirchner T, Müller-Hermelink HK. Characterization of a protein with an acetylcholine receptor epitope from myasthenia gravis associated thymomas. Lab Invest 1990;62:279-86.
12 Gilhus NE, Willcox N, Harcourt G, Nagvekar N, Beeson D, Vincent A, Newsom-Davis J. Antigen-presentation by thymoma epithelial cells from myasthenia gravis patients to potentially pathogenic T cells. J Neuroimmunol 1995;56:65-76.
13 Fujinami RS, Oldstone MBA. Amino acid homology between the encephalitogenic site of myelin basic protein and virus: mechanism for autoimmunity. Science 1985;230:1043-5.
14 Wraith DC, McDevitt HO, Steinman L, Acha-Orbea H. T cell recognition as the target for intervention in autoimmune disease. Cell 1989;57:709-15.
15 Hohlfeld R, Toyka KV, Heininger K, Grosse-Wilde H, Kalies I. Autoimmune human T lymphocytes specific for acetylcholine receptor. Nature 1984;310: 244-6.
16 Hohlfeld R, Kalies I, Kohleisen B, Heininger K, Conti-Tronconi B, Toyka KV. Myasthenia gravis: Stimulation of antireceptor autoantibodies by autoreactive T cell lines. Neurology 1986;36:618-21.
17 Sun JB, Harcourt G, Wang ZY, Hawke S, Olsson T, Fredrikson S, Link H. T cell responses to human recombinant acetylcholine receptor α-subunit in

myasthenia gravis and controls. Eur J Immunol 1992;22:1553-9.

18 Melms A, Chrestel S, Schalke BCG, Wekerle H, Mauron A, Ballivet M, Barkas T. Autoimmune T lymphocytes in myasthenia gravis. Determination of target epitopes using T lines and recombinant products of the mouse nicotinic acetylcholine receptor gene. J Clin Invest 1989;83:785-90.

19 Ong B, Willcox N, Wordsworth P, Beeson D, Vincent A, Altmann D, Lanchbury JSS, Harcourt GC, Bell JI, Newsom-Davis J. Critical role for the gly/val 86 dimorphism in autoantigen presentation to human T cells. Proc Natl Acad Sci USA 1991;88:7343-7.

20 Melms A, Malcherek G, Gern U, Wiethölter H, Müller CA, Schöpfer R, Lindstrom J. T cells from normal and myasthenic individuals recognize the human acetylcholine receptor: heterogeneity of antigenic sites on the α-subunit. Ann Neurol 1992;31:311-8.

21 Sommer N, Harcourt GC, Willcox N, Beeson D, Newsom-Davis J. Acetylcholine receptor-reactive T lymphocytes from healthy subjects and myasthenia gravis patients. Neurology 1991;41:1270-6.

22 Link H, Olsson O, Sun JB, Wang WZ, Andersson G, Ekre HP, Brenner T, Abramsky O, Olsson T. Acetylcholine receptor reactive T and B cells in myasthenia gravis and controls. J Clin Invest 1991;87:2191-6.

23 Germain RN, Margulies DH. The biochemistry and cell biology of antigen processing and presentation. Annu Rev Immunol 1993;11:403-50.

24 Protti MP, Manfredi A, Horton RM, Bellone M, Conti-Tronconi BM. Myasthenia gravis: recognition of a human autoantigen at the molecular level. Immunol Today 1993;14:363-8.

25 Cresswell P. Assembly, transport, and function of MHC class II molecules. Annu Rev Immunol 1994;12:259-93.

26 Compston DAS, Vincent A, Newsom Davis J, Batchelor JR. Clinical, pathological, HLA antigen and immunological evidence for disease heterogeneity in myasthenia gravis. Brain 1980;103:579-601.

27 Feltkamp TEW, Van den Berg-Loonen PM, Nijenhuis LE, Engelfriet CP, Van Rossum AL, Van Loghem JJ, Oosterhuis HJGH. Myasthenia gravis, autoantibodies and HL-A antigens. Br Med J 1974;i:131-3.

28 Babbitt B, Allen PM, Matsueda G, Haber E, Unanue ER. Binding of immunogenic peptides to Ia histocompatibility molecules. Nature 1985;317:359-61.

29 Stern LJ, Brown JH, Jardetzky TS, Gorga JC, Urban RG, Strominger JL, Wiley DC. Crystal structure of the human class II MHC protein HLA-DR1 complexed with an influenza virus peptide. Nature 1994;368:215-21.

30 Rammensee HG, Friede T, Stevanovic S. MHC ligands and peptide motifs: first listing. Immunogenetics 1995;41:178-228.

31 Christadoss P, Lennon VA, David C. Genetic control of experimental autoimmune myasthenia gravis in mice. I. Lymphocyte proliferative responses to acetylcholine receptor is under H-2 linked Ir gene control. J Immunol 1979;123:2540-3.

32 Bellone M, Ostile N, Lei S, Wu XD, Conti-Tronconi B. The I-A^{bm12} mutation, which confers resistence to experimental autoimmune myasthenia gravis, drastically affects the epitope repertoire of murine CD4+ cells sensitized to nicotinic acetylcholine receptor. J Immunol 1991;147:1484-91.

33 Falk K, Rötzschke O, Stevanovic S, Jung G, Rammensee HG. Allele-specific motifs revealed by sequencing of self-peptides eluted from MHC molecules. Nature 1991;351:290-6.
34 Malcherek G, Falk K, Rötzschke O, Rammensee H-G, Stevanovic S, Gnau V, Jung G, Melms A. Peptide motif of two HLA molecules associated with myasthenia gravis. Int Immunol 1993;5:1229-37.
35 Malcherek G, Gnau V, Stevanovic S, Rammensee HG, Jung G, Melms A. Characterization of allele-specion of allele-specific contact sites of DR17 ligands. J Immunol 1994;154:1141-9.
36 Geluk A, Van Meijgaarden KE, Janson AAM, Wouter Drijfhout J, Meloen RH, De Vries RR, Ottenhoff THM. Functional analysis of DR17 (DR3)-restricted mycobacterial T cell epitopes reveals DR17 binding motif and enable the design of allele-specific competitor peptides. J Immunol 1992;149:2864-71.
37 Hohlfeld R, Toyka KV, Tzartos SJ, Carson W, Conti-Tronconi B. Human T helper lymphocytes in myasthenia gravis recognize the nicotinic receptor alpha subunit. Proc Natl Acad Sci USA 1987;84:5379-83.
38 Fuji Y, Lindstrom J. Specificity of the T cell response to acetylcholine receptor in experimental autoimmune myasthenia gravis. Response to subunits and synthetic peptides. J Immunol 1988;140:1830-7.
39 Brocke S, Brautbar C, Steinman L, Abramsky O, Rothbard J, Neumann D, Fuchs S, Mozes E. In vitro proliferative responses and antibody titers specific to human acetylcholine receptor synthetic peptides in patients with myasthenia gravis and relation to HLA class II genes. J Clin Invest 1988;82:1894-1900.
40 Matsuo H, Batocchi A-P, Hawke S, Nicolle M, Jacobson L, Vincent A, Newsom-Davis J, Willcox, N. Peptide-selected T cell lines from myasthenia gravis patients and controls recognize epitopes that are not processed from whole acetylcholine receptor. J Immunol 1995;155:3683-92.
41 Schild HJ, Rötzschke O, Kalbacher H, Rammensee HG. Limit of T cell tolerance to self proteins by peptide presentation. Science 1990;247:1587-9.
42 Lehmann PV, Forsthuber T, Miller A, Sercarz EE. Spreading of T cell autoimmunity to cryptic determinants of an autoantigen. Nature 1992;358:155-7.
43 Lanzavecchia A. How can cryptic epitopes trigger autoimmunity? J Exp Med 1995;181:1945-8.
44 Gammon G, Sercarz E. How some T cells escape tolerance induction. Nature 1989;342:183-5.
45 Liu GY, Fairchild PJ, Smith RM, Prowle JR, Kioussis D, Wraith DC. Low avidity recognition of self-antigen by T cells permits escape from central tolerance. Immunity 1995;3:407-15.

KARINE MIGNON-GODEFROY, LILA TAKORABET, ORTWIN ROTT
AND JEANNINE CHARREIRE

Role of pharmacological substances in the regulation of Experimental Autoimmune Thyroiditis (EAT)

Introduction

Experimental Autoimmune Thyroiditis (EAT) is a transient organ-specific autoimmune disease which develops in genetically susceptible strains of mice[1] approximately one month following immunization with either autologous[2] or xenogeneic thyroglobulin (Tg) and adjuvants[3] or with human Tg peptides emulsified in adjuvants.[4-7] When EAT is induced by Tg and adjuvants, H-2^k or H-2^s mice develop EAT, whereas H-2^b mice do not.[1] Furthermore, EAT is also inducible in irradiated susceptible recipients by injection of thyroid-[8] or Tg-[9,10] specific T cells. EAT is characterized by circulating anti-Tg autoantibodies and infiltrations of the thyroid gland by lymphoid cells, including CD8+ T cells.[11,12] Remarkably, autoantibodies to Tg are detectable at high levels when EAT is induced by Tg, whereas they are at low levels when EAT is induced by Tg-specific T cells.[8,10]

Peptides inducing EAT

The last few years, various laboratories, including our own, have determined peptide sequences able to induce EAT within the huge Tg molecule or in the thyroid peroxydase (TPO) molecule (see Table 1). Most of the peptides were selected for their ability to stimulate TPO- or Tg-specific T cell lines or clones. The only EAT-inducing peptide from TPO comprised 15 amino-acids located in the NH_2-terminal part of porcine TPO.[13] In contrast to Tg EAT-inducing peptides, TPO peptide induces EAT in H-2^b mice but never in H-2^k animals.

Several peptides were selected from Tg from various species. An

TABLE 1 EAT-inducing peptides.

Protein	genetic control	localization in the protein	Authors
porcine TPO	H-2^b	744-788	Kotani (1990)
human Tg	H-2^k	T4 in position 2553	Champion(1991)
human Tg	H-2^k	1672-1711	Texier (1992)
rat Tg	H-2^k	2495-2511	Chronopoulou (1992)
rat Tg	H-2^k	2695-2713	Carayanniotis (1994)

11 amino-acid peptide from human Tg containing the thyroxine peptide (T4) located in position 2553 in the Tg molecule appeared to be involved in EAT induction.[14] It must be noted that this peptide alone is unable to induce EAT, whereas a T cell clone specific for this peptide can do so.

Another peptide which induces EAT in CBA/J mice was derived from human Tg on its ability to be recognized by one Tg-specific cytotoxic T cell hybridoma. It is a 40 amino-acid long, non-iodinated tryptic fragment located in the middle (amino-acids 1672-1711) of the human Tg molecule. We named this peptide F40D.[7]

Simultaneously, another non-iodinated 17 amino-acid long peptide from rat Tg, corresponding to amino-acids 2495 to 2511 from human Tg, was selected. The authors demonstrated that it induced EAT in the same strains of mice as murine or xenogeneic Tg.[4] Recently, the latter team reported that another non-iodinated 18 amino-acid long peptide from human Tg, located in positions 2695 to 2713 of Tg, could also induce EAT in the EAT-susceptible H-2^s mice, whereas it was ineffective in H-2^k mice.[5]

On the whole, these data demonstrate that at least two thyroid antigens, Tg and TPO, are involved in EAT induction and that in the huge Tg molecule different peptides can induce EAT. Iodination is not a prerequisite for induction of EAT by a Tg peptide.

Role of recombinant human IL-10 (rhIL-10) in EAT

The role of T cells in EAT was first demonstrated by the proliferative response to Tg and the induction of the disease by Tg-specific T cell clones. It is now well known that T cells can also act through the cytokines they secrete.[15]

We have previously shown that EAT could be induced in susceptible strains of mice by intrathyroidal injection of IFN-γ[16] and that a neutralizing mAb to IFN-γ could prevent EAT induction by decreasing the number of Tg-specific T cells.[17] More recently, in a model of transfer-EAT, we demonstrated that *in vitro* treatment of murine Tg (mTg)-activated T cells with rhIL-10 reduced their proliferative and cytotoxic responses to mTg and significantly decreased their capacity to induce EAT after transfer into irradiated syngeneic recipients.[18] We postulated that the inhibition of mTg-specific T cell function in the presence of rhIL-10 might result from a decrease in mTg presentation by APCs, and/or from the inhibition of the proinflammatory cytokine production, such as IL-1 or TNF-α. Interestingly, in the same series of experiments, another 'Th$_2$-typical' cytokine, IL-4, was completely ineffective in decreasing the mTg-specific T cell functions or exerting a beneficial effect on the subsequent development of transfer-EAT.[18]

Those experiments led us to investigate whether beneficial effects could be achieved *in vivo* by IL-10 treatment of CBA/J mice with EAT induced by mTg and adjuvants. In these experiments, rhIL-10 was given to mice in both preventive and curative protocols.

PREVENTIVE PROTOCOL In a first set of experiments, treatment with i.p. rhIL-10 was begun 24 hours after primary immunization and restarted 15 days after challenge. Three dosage levels of rhIL-10 were administered: 600, 60 and 6 ng per injection. Compared to the proliferative response of spleen cells from mTg-immunized mice to mTg, the proliferative responses of treated mice were totally abolished. They were similar to those observed in spleen cells from control mice that received only adjuvant suspensions and were not immunized. The cytotoxic responses towards mTg-pulsed syngeneic macrophages were also abolished in lymph node cells from IL-10 treated mice. It should be noted that this effect was most clear-cut on day 28 after immunization, when the cytotoxic response was the highest (see Table 2).

Lymphocytic infiltrations in the thyroid glands were evaluated in the treated and control mice, injected with IL-10 or not. As expected, when animals were not immunized or received only adjuvant suspensions, they did not develop lymphocytic infiltrations in their thyroid glands. In contrast, when they were immunized with mTg they developed EAT as assessed by a mean infiltration index of

TABLE 2 Preventive effect of rhIL-10 on EAT development.

	MTg-immunized mice treated with rhIL-10*(ng)			
	0	6 × 600	6 × 60	6 × 6
Proliferative response to mTg** (day 21 post mTg)	Maximal	abolished	abolished	N.D***.
Cytotoxic response to mTg** (day 28 post mTg)	Maximal	abolished	abolished	abolished
Autoantibodies to mTg	Maximal	Maximal	Maximal	Maximal
Lymphocytic infiltration in the thyroid gland****	2.11±0.13	1.37±0.32	1.18±0.25	2.08±0.37
Cell subsets in the spleen cells	normal	decreases in CD4+, CD8+, B220+, Class II	N.D.	N.D.

*: rhIL-10 was administered on days 1, 3, and 5 after primary immunization by mTg and adjuvants and on days 15, 17 and 19 after challenge by mTg.

**: simultaneously tested proliferative and cytotoxic responses towards cytochrome C were not modified by rhIL-10 treatment.

***: N.D. not done.

****: expressed as indexes.

2.11. When mice were injected 6 times with 600 or 60 ng rhIL-10, significant reductions in lymphocytic infiltrations (46 and 57%, respectively) were observed. Lastly, when mice were injected with 6 × 6 ng rhIL-10, no reduction in lymphocytic infiltrations was observed, the mean infiltration index being comparable to that found in mTg-immunized mice.

CURATIVE PROTOCOL The significant prevention of EAT by rhIL-10 led us to study whether this cytokine could act once the disease had started, i.e. when autoantibodies to mTg were found in sera and when proliferative and cytotoxic responses were detectable in lymphoid organs. Therefore, rhIL-10 was injected only at the time of challenge with mTg and incomplete adjuvant.

The proliferative and cytotoxic responses to mTg were evaluated on days 21 and 28 after mTg immunization and compared to those observed in adjuvant-treated mice that had been injected with neither mTg nor IL-10 and in mTg-immunized mice that had not been treated with IL-10. The proliferative responses to mTg were partially abolished in spleen cells from IL-10 treated mice. A proliferative response comparable to that in control animals could only be achieved using the highest concentration of mTg during *in vitro* stimulation of spleen cells and the highest amount of IL-10. The cytotoxic responses towards mTg-pulsed syngeneic macrophages were also abolished in lymph node cells of IL-10 treated mice. As reported in the preventive protocol, this effect was most clear-cut on day 28 after immunization, when the cytotoxic response was the highest. It should also be noted that on day 28 after mTg immunization the effect of rhIL-10 was dose-dependent: the higher the amount of rhIL-10 injected, the greater the inhibition of the mTg-specific cytotoxic response (see Table 3).

As previously shown, when animals had not been immunized or had only received adjuvant suspensions, they did not develop lymphocytic infiltrations in their thyroid glands. When they had been immunized with mTg they developed EAT, as assessed by a mean infiltration index of 2.1. When mice had been injected 3 times with 600 ng rhIL-10, a significant reduction (around 44%) in the mean lymphocytic infiltration index was observed. No significant effect was observed with 60 or 6 ng rhIL-10 on lymphocytic infiltration of the thyroid glands; however, there were reductions of 32 and 22% in lymphocytic infiltration.

To understand the mechanism(s) responsible for EAT suppression by rhIL-10, cell subsets were determined in non-adherent lymphoblastoid spleen cells from rhIL-10 treated mice. We found that both CD4 and CD8 cells were decreased by approximately 45%, and that B and Ia+ cells were decreased by approximately 20%. Hence, we then investigated whether apoptosis could be responsible for these cell decreases. To this end, purified T cells from mice sacrificed on day 21 or 28 after immunization with mTg were analyzed for signs of DNA fragmentation. In both the preventive and curative protocols, an increased percentage of apoptotic T lymphocytes was found in both groups on day 21 after immunization. This increase in apoptotic cells was directly correlated with the amount of rhIL-10 injected into the animals and more pro-

TABLE 3 Curative effect of rhIL-10 after EAT development

	MTg-immunized mice treated with rhIL-10* (ng)			
	0	3 x 600	3 x 60	3 x 6
Proliferative response to mTg (day 21 post mTg)	Maximal	reduced by 75%	reduced by 50%	N.D**.
Cytotoxic response to mTg (day 28 post mTg)	Maximal	reduced by 95%	reduced by 75%	reduced by 35%
Autoantibodies to mTg		Maximal	Maximal	Maximal
Lymphocytic infiltration in the thyroid gland***	2.11±0.13	1.40±0.28	1.58±0.29	1.75±0.42
Cell subsets in the spleen cells	Normal	decreases in CD4+, CD8+, B220+, Class II	N.D.	N.D.

*: rhIL-10 was administered on days 15, 17 and 19 after challenge with mTg.
**: N.D.= not done.
***: expressed as indexes.

nounced on day 21 than on day 28.

These data led us to conclude that *in vivo* administration of a high dose of rhIL-10 interferes with the process of activation-induced cell death in the mature T lymphocyte compartment.

Role of phenothiazines in the induction of thyroid autoimmune disorder

It has been reported that two phenothiazines: alimemazine and chlorpromazine, could induce both a decrease in thyroid hormone levels and hyperreactivity of the thyroid gland when given orally to rats. These effects of phenothiazines led us to investigate whether one particular phenothiazine, alimemazine, currently used in psychiatry and pediatrics, could affect MHC antigen expression on thyroid cells, and thus immune activity.

In a first set of experiments, the effects of alimemazine on MHC class I and class II antigen expression were evaluated by FACS analysis on a human thyroid epithelial cell line (TEC): GEJ,[19] 24, 48 and 72 hours after addition of 10^{-8}, 10^{-7} or 10^{-6} M alimemazine. No significant effect on MHC class I antigen expression was observed. In contrast, MHC class II antigen expression increased significantly 24 and 48 hours after the addition of alimemazine to human TEC. This effect was transient, since the increases in MHC class II antigens were no longer significant 72 hrs after addition of alimemazine. It must be noted that when the same experiment was performed on B, T and NK cell lines, no modification in MHC class II antigen expression was observed after addition of 10^{-6} M alimemazine.

Clear increases in surface expression of MHC class II antigen were observed on human GEJ cells 24 and 48 hours after incubation with alimemazine. We further investigated, by Northern blot analysis, the kinetics of the induction of MHC class I, II and Ii chain transcripts by alimemazine in both GEJ cells and primary cultures of human and murine thyroid epithelial cells. When GEJ cells were incubated with alimemazine, maximal levels of class II and Ii chain transcripts were transiently observed at 5 or 2 hrs, respectively, followed by a return to baseline levels by 24 hrs. These increases in MHC class II and Ii chain transcripts were observed with all three concentrations of alimemazine used, whereas no alteration in the level of MHC class I mRNA was observed at any time or with any concentration of alimemazine studied (see Table 4).

In order to test whether MHC class II antigens induced by alimemazine on thyroid cells could play a functional role, syngeneic mTg-specific and allogeneic stimulations were performed. Lymph node cells from syngeneic mice, previously primed and challenged with mTg, as well as from naive allogeneic mice were added to primary cultures of thyroid epithelial cells in which MHC class II antigens were previously induced by addition, 48 hrs earlier, of 10^{-7} M alimemazine. As a positive control, aliquot cells were deposited onto murine thyroid epithelial cells in which class II antigens were induced by IFN-γ. Syngeneic mTg-specific and allogeneic proliferative responses were observed when murine thyroid epithelial cells were previously incubated with recombinant rat IFN-γ or with 10^{-7} M alimemazine. More precisely, when alimemazine and IFN-γ were added to murine thyroid epithelial cells, increases in mTg-specific and allogeneic proliferative responses were much greater with alimemazine than with IFN-γ. Lastly, the syngeneic mTg-specific and allogeneic proliferative responses which were observed at the three cell dilutions tested were inhibited when anti-murine class II mAb was added to murine thyroid epithelial cells before lymph node

TABLE 4 Alimemazine (10^{-7} M) induces, 48 hrs after its addition, *de novo* MHC class II antigen expression in Thyroid Epithelial Cells (TEC).

	Class I		Class II		Ii chain	
	Protein	mRNA	Protein	mRNA	Protein	mRNA
GEJ cells*	not modified	not modified	+	+++	N.D.	++
Primary cultures of human TEC*	N.D.	not modified	N.D	+++	N.D.	++
Primary cultures of murine TEC*	N.D.	not modified	N.D.	+++	N.D.	++

*: In the absence of alimemazine only MHC class I protein and mRNA are detectable.

cell addition, whereas control mAb was ineffective.[20]

Multiple and complex mechanisms are involved in the induction and development of autoimmune thyroid disorders. Whereas some defined Tg-specific T cells can induce EAT, the same T cells can, if inactivated, also prevent the development of EAT. Furthermore, T cells can also act through the cytokines they produce and either induce EAT (such as IFN-γ), prevent or still more interestingly cure the disease (this was demonstrated with IL-10). Our most recent data,[21] demonstrating that some particular subsets of autoantibodies or drugs induce MHC Class II antigens on the TEC surface, transforming these cells into APC, bring new insights into their role. The transformation of a TEC into an APC, hypothesized due to the previously described induction of class II antigen on TEC,[21-23] has now been shown under physiological conditions.

References

1 Vladutiu AO, Rose NR. Transfer of experimental autoimmune thyroiditis of the mouse by serum. J Immunol 1971; 106:1139.

2 Rose NR, Witebsky E. Studies on organ specificity. V. Changes in the thyroid glands of rabbits following active immunization with rabbit thyroid extracts. J Immunol 1956; 76:417.

3 Charreire J. Immune mechanisms in autoimmune thyroiditis. Adv Immunol 1989; 46:263.

4 Chronopoulou E, Carayanniotis G. Identification of a thyroiditogenic sequence

within the thyroglobulin molecule. J Immunol 1992; 149:1039.
5 Carayanniotis G. Chronopoulou E, Rao VP. Distinct genetic pattern of mouse susceptibility to thyroiditis induced by a novel thyroglobulin peptide. Immunogenetics 1994; 39:21.
6 Salamero J, Remy JJ, Michel-Béchet M, Charreire J. Experimental autoimmune thyroiditis induced by a 5-10 kDa tryptic fragment from porcine thyroglobulin. Eur J Immunol 1987; 17:843.
7 Texier B, Bédin C, Tang H, Camoin L, Laurent-Winter C, Charreire J. Characterization and sequencing of a 40-amino-acid peptide from human thyroglobulin inducing experimental autoimmune thyroiditis. J Immunol 1992; 148:3405.
8 Charreire J, Michel-Béchet M. Syngeneic sensitization of mouse lymphocytes on monolayers of thyroid epithelial cells. III. Induction of thyroiditis by thyroid-sensitized T lymphoblasts. Eur J Immunol 1982; 12:421.
9 Maron R, Zerubavel R, Friedman A, Cohen IR. T lymphocyte line specific for thyroglobulin produces or vaccinates against autoimmune thyroiditis in mice. J Immunol 1983; 131:2316.
10 Romball CG, Weigle WO. T cell competence to heterologous and homologous thyroglobulins during the induction of experimental autoimmune thyroiditis. Eur J Immunol 1984; 14:887.
11 Creemers P, Rose NR, Kong YCM. Experimental autoimmune thyroiditis. *In vitro* cytotoxic effects of T lymphocytes on thyroid monolayers. J Exp Med 1983; 157:559.
12 Salamero J, Charreire J. Syngeneic sensitization of mouse lymphocytes on monolayers of thyroid epithelial cells. VII. Generation of thyroid-specific cytotoxic effector cells. Cell Immunol 1985; 91:111.
13 Kotani T, Umeki K, Hirai K, Ohtaki S. Experimental murine thyroiditis induced by porcine thyroid peroxidase and its transfer by the antigen-specific T cell line. Clin Exp Immunol 1990; 80:11.
14 Champion BR, Page KR, Parish N, Rayner DC, Dawe K, Biswas-Hughes G, Cooke A, Geysen M, Roitt IM. Identification of a thyroxine-containing self-epitope of thyroglobulin which triggers thyroid autoreactive T cells. J Exp Med 1991; 174:363.
15 Paul WE, Seder RA. Lymphocyte responses and cytokines. Cell 1994; 76:241.
16 Rémy JJ, Salamero J, Michel-Béchet M, Charreire J. Experimental autoimmune thyroiditis induced by recombinant interferon-γ. Immunol Today 1987; 8:73.
17 Tang H, Mignon-Godefroy K, Meroni PL, Garotta G, Charreire J, Nicoletti F. The effects of a monoclonal antibody to interferon-γ on experimental autoimmune thyroiditis (EAT): prevention of disease and decrease of EAT-specific T cells. Eur J Immunol 1993; 23:275.
18 Mignon-Godefroy K, Brazillet MP, Rott O, Charreire J. Distinctive modulation by IL-4 and IL-10 of the effector function of murine thyroglobulin-primed cells in 'transfer-experimental autoimmune thyroiditis'. Cell Immunol 1995;195:171.
19 Karsenty G, Michel-Béchet M, Charreire J. Monoclonal human thyroid cell line GEJ expressing human thyrotropin receptors. Proc Natl Acad Sci USA

1985; 82:2120.
20 Takorabet L, Ropars A, Raby C, Charreire J. Phenothiazine induces 'de novo' MHC class II antigen expression on thyroid epithelial cells: A new mechanism for drug-induced autoimmunity. J Immunol 1995;154:3593.
21 Bottazzo GF, Pujol-Borrell R, Hanafusa T, Feldmann M. Role of aberrant HLA-DR expression and antigen presentation in induction of endocrine autoimmunity. Lancet 1983; 2:1115.
22 Salamero J, Michel-Béchet M, Charreire J. Différences de répartition des antigènes de classe I et II du complexe majeur d'histocompatibilité (CMH) à la surface des cellules épithéliales de thyroïde (CET) murines en culture. CR Acad Sci III 1981; 293:745.
23 Salamero J, Michel-Béchet M, Wietzerbin J, Charreire J. Syngeneic sensitization of mouse lymphocytes on monolayers of thyroid epithelial cells (TEC). VIII: γ interferon induced Ia antigen expression on TEC cultures. Tissue Antigens 1985; 25:266.

GERD PLUSCHKE

Immunotherapy in autoimmune diseases

Introduction

An ideal immunotherapy would correct altered immune mechanisms and restore immune homeostasis. However, although impaired immunological regulation seems to be a common feature of autoimmune diseases, their treatment still remains essentially symptomatic. Autoimmune diseases comprise a complex and heterogenous category of disorders and since different immune mechanisms and various autoantigens may be of importance during the distinct stages of a particular autoimmune disease, it will never be easy to design a form of therapeutic immunosuppression that can be given over a short-term period and that will result in long-term therapeutic effects without impairing the body's defense against infectious agents. Current immunosuppressive therapeutic regimens (Table 1) are relatively unspecific[1] and often associated with severe side effects which are not acceptable for those forms of autoimmune disease that are not life threatening.

It has been suggested that a common feature of most human autoimmune diseases is the activation of autoreactive T cells. T cell activation requires interaction among the components of the ternary complex formed by a T cell receptor (TCR) for antigen, a MHC gene product and a peptide antigen presented by the MHC gene product. New concepts of a more specific immunotherapy have therefore targeted the components of the ternary complex, each of which may serve as an independent target for specific immunotherapy.[2,3] Implementation of a truly selective immunosuppression may, however, depend on identification of the autoantigens involved. Different strategies of selective immunointervention have been shown to be effective in animal models of autoimmune disease (Ta-

Table 1. Therapy of autoimmune diseases

Non-specific
Corticosteroids
Cytostatic agents
Gold salts
D-Penicillamine
Levamisole
Thymic hormones
Apharesis
Partially specific
Cyclosporin A
Thoracic duct drainage
Total lymphoid irradiation
Intravenous immunoglobulins
Antibodies against the T cell antigens CD4, CD5, CD7 and CDw52
Targeting of the interleukin-2 receptor
Targeting of adhesion molecules
Targeting of TNFa and other cytokines
Specific
T cell vaccination and T cell receptor targeted immunotherapy
Targeting of the peptide-MHC complex
Tolerance induction by oral administration of antigen

ble 2), where the autoimmune pathology is induced by treatment with a known autoantigen. However, only a few candidate autoantigens have so far been identified in human autoimmune diseases. These include myelin basic protein (MBP) in multiple sclerosis (MS), thyroid peroxidase in Grave's disease and the acetylcholine receptor in myasthenia gravis. Furthermore, there is mounting evidence that the expressed autoimmune repertoire is not constant during the development of an autoimmune disease.[4] While an autoimmune response may initially be limited to a single antigenic determinant of an autoantigen,[5] both intramolecular and intermolecular determinant-spreading may occur as the disease progresses and lymphocytes that react to a variety of autoantigenic determinants may be involved in disease progression. The development of truly

TABLE 2 Strategies for selective immunosuppressive therapy

Target cells	Strategy	Therapeutic agents	Information required	Example
Antigen-presenting cells	general blockade of relevant MHC	MHC antagonists	MHC gene products involved in the presentation of autoantigenic epitopes	Guéry et al.[54]
		anti-MHC antibodies		Steinman et al.[55]
	blockade of complexes of MHC and autoantigenic epitopes	antibodies against MHC-peptide complexes	autoantigenic epitopes and MHC gene products involved in their presentation	Aharoni et al.[36]
Autoreactive T cells	T cell receptor antagonism	analogues of autoantigenic epitopes	autoantigenic epitopes	Alexander et al.[56]; De Magistris et al.[57]
	depletion of T cells that express relevant T cell receptor V regions	T cell receptor V region specific antibodies	T cell receptor V region usage of disease-mediating T cells	Heber-Katz & Acha-Orbea[26]; Zaller et al.[35]
	tolerance induction	autoantigenic epitope in tolerogenic formulation	autoantigenic epitopes	Clayton et al.[58]; Gaur et al.[59]
		soluble complexes of MHC and autoantigenic epitopes	autoantigenic epitopes and MHC gene products involved in their presentation	Sharma et al.[60]
Regulatory cells	induction of regulatory cells by T cell vaccination	attenuated disease-mediating T cell	access to disease-mediating T cells	Cohen[61]
	induction of T cell receptor specific regulatory cells	T cell receptor V region derived peptides	T cell receptor V region usage of disease-mediating T cells	Howell et al.[33]; Vandenbark et al.[34]
	induction of regulatory cells by oral administration of autoantigen	autoantigen	autoantigen	Zhang et al.[62] Brod et al.[63]

selective immunotherapeutic strategies which will lead to the inactivation of only the autoreactive lymphocyte clones is still severely hampered by lack of knowledge of the pathogenic events and autoantigens involved in disease initiation and progression.

One approach to antigen-specific immunotherapy is to block recognition of the autoantigen(s) by autoreactive T cells (Table 2). This can only work when the autoimmune response is restricted in the number of epitopes recognized or in the repertoire of TCRs used. The second therapeutic strategy involves the antigen-specific targeting of T cells that down-regulate immune responses at the site of inflammation (Table 2). Here it may not be necessary that the antigen which has initiated the autoimmune process is known, nor that the disease-inducing T cells are restricted in terms of epitopes recognized or TCRs used. One example for this second approach is oral tolerization, where it is hoped that regulatory T cells will be elicited which will migrate to the site of inflammation and locally suppress immune responses.

Nonspecific immunotherapy

Corticosteroids, which affect many of the steps involved in inflammation, are commonly used in the therapy of human autoimmune diseases. Cytostatic agents can also work as nonspecific immunosuppressants.[6] Among these, methotrexate, azathioprine and cyclophosphamide are the most commonly used. Certain slow-acting drugs, such as levamisole, gold salts and D-penicillamine have been shown to have beneficial effects in short-term treatment.[1] Beneficial effects have also been documented for treatment regimens with thymic hormones.[7] Apharesis has been used to remove blood components (immunoglobulins, immune complexes and cytokines) from the intravascular compartment of patients with autoimmune diseases, and beneficial effects have been obtained in myasthenia gravis, MS, rheumatoid vasculitis and Guillain-Barré syndrome.[1]

Partially specific immunotherapy

THORACIC DUCT DRAINAGE, TOTAL LYMPHOID IRRADIATION, CYCLOSPORIN AND INTRAVENOUS IMMUNOGLOBULINS Thoracic duct drainage and total lymphoid irradiation induce persistent immunosuppression, and results have been reported from the treatment of a number of autoimmune diseases.[1] Treatment with cyclosporin, which directly affects T cells, has been shown to have beneficial effects in various autoimmune diseases.[8] However, these therapeutic effects seem to be dependent on continuous drug administration, with the risk of severe side effects. Intravenous immunoglobulins have been used in various autoimmune diseases with encouraging results.[1] It has been suggested that these preparations of polyreactive IgG contain anti-idiotypes against idiotypes expressed by disease-associated B cells, and that they thus interfere with the function of the autoimmune network.[9]

MONOCLONAL ANTIBODIES AGAINST THE T CELL ANTIGENS CD4, CD5, CD7 AND CDW52 A variety of monoclonal antibodies with specificity for proteins expressed on the cell surface of human T cells have been used for immunointervention.[10] In clinical trials with anti-CD5 antibody-toxin conjugates and anti-CD4, anti-CD7 and anti-CDw52 antibodies, pronounced decreases in the number of circulating T cells were observed. In animal models it has been demonstrated that antibodies directed against certain adhesion receptors on T cells are capable of establishing long-term tolerance.[11] In this connection, the results of a study in patients with severe systemic vasculitis[12] are encouraging. Using the humanized antibody CAMPATH-1H (anti-CDw52), with or without additional anti-CD4 antibodies, long-term remissions have been achieved in some of the patients. However, the significance of the responses described so far for a number of autoimmune diseases, including rheumatoid arthritis (RA), MS, psoriasis, and inflammatory bowel disease, is difficult to evaluate, because data from randomized, double-blind, placebo-controlled trials are still largely missing. The results of the first controlled trials with anti-CD4 antibodies and anti-CD5 immunoconjugates were disappointing.[13-15]

TARGETING OF THE INTERLEUKIN-2 RECEPTOR Since the high affinity interleukin-2 receptor is only transiently expressed during antigen-triggered activation of T lymphocytes, it was thought that the targeting of this receptor would preferentially affect those cells which are activated in the autoimmune process.[16] Clinical trials with antibodies against the alpha chain of the IL-2 receptor (CD25) and with interleukin-2-diphtheria toxin fusion protein are under way. The first results of a controlled trial with the fusion protein in RA were not encouraging.[17]

TARGETING OF ADHESION MOLECULES Since leukocyte adhesion to endothelial cells and extravasation into the site of inflammation are thought to be key steps in autoimmune diseases, there is increasing clinical interest in the specific inhibition of these processes.[18] Because of its multistep nature, the process of recruitment of inflammatory cells (cell activation, leukocyte rolling, firm adhesion, and transendothelial migration) presents a variety of potential therapeutic targets. These include 1) adhesion molecules on the endothelial surface, such as E-selectin, vascular cell adhesion molecule (VCAM)-1 and intercellular adhesion molecule (ICAM)-1, and 2) ligands for these molecules on the inflammatory cells such as E-selectin ligand, very late activation antigen (VLA)-4 and CD11/18. Several approaches to the blockade of the adhesion cascade, including monoclonal antibodies, soluble oligosaccharides and polypeptides, have demonstrated efficacy in animal models.[18] Since multiple adhesion pathways exist, targeting of a single pathway may, however, not be sufficient in a clinical setting.

TARGETING OF TUMOR NECROSIS FACTOR (TNF) α AND OTHER CYTOKINES Cytokines such as TNFα, IFN-γ and many interleukins are found in the sites of inflammation in autoimmune diseases. They produce a variety of effects, some of which result in pathological changes. Studies on the cytokine network have indicated that TNFα plays a key role in autoimmune disease progression.[19] These investigations suggested that neutralization of TNFα with antibodies would significantly inhibit the generation of other pro-inflammatory cytokines. Data from a placebo-controlled clinical trial in RA patients with two intravenous doses of the chimeric anti-TNFα antibody cA2 support the concept that TNFα is implicated in the pathogenesis of RA. Four weeks post infusion, only 8% of the patients receiving placebo improved, compared with 44% of those receiving 1 mg/kg cA2 and 79% of those receiving 10 mg/kg.[19] Encouraging results have also been reported from clinical trials in RA with a humanized murine anti-TNFα antibody[20] and with a fusion protein comprising recombinant human p75 TNF receptor coupled to a human IgG1 Fc framework region.[21] In addition to TNFα, IL-1 and IL-6 are also targets of anti-cytokine based therapeutic strategies currently being tested in clinical trials (Table 3).

TABLE 3 Anti-cytokine based therapies in clinical trials in RA

Target	Therapeutic agent	Company
TNFα	Monoclonal anti-TNFα antibodies:	
	- chimeric (cA2)	Centocor
	- humanized (CDP571)	CellTech
	Soluble TNF receptor-fusion proteins:	
	- TNF receptor:Fc (p55)	Roche
	- TNF receptor:Fc (p75)	Immunex
IL-1	IL-1 receptor antagonist	Synergen
	Soluble IL-1 receptors	Immunex
IL-6	Monoclonal anti-IL-6 antibodies:	
	- murine (BE-8)	

(adapted from Maini et al.[19])

Specific immunotherapy

T CELL VACCINATION AND T CELL RECEPTOR TARGETED IMMUNOTHERAPY In animal models, the development of autoimmune diseases can be suppressed by injecting attenuated autoantigen-reactive T cells.[22] Anti-idiotypic or anti-clonotypic immune responses may be responsible for the immunotherapeutic effects, by affecting the pathogenic T cells in the host. T cell vaccination in patients with MS, using a vaccine prepared from irradiated myelin basic protein (MBP)-reactive T cells resulted in a decrease in the frequency of circulating MBP-reactive T cells.[23,24] Magnetic resonance imaging showed a mean 8% increase in brain lesions in the vaccinated patients, compared with 39.5% increase in control patients.[23] In another clinical trial, patients with RA were vaccinated with synovial T cells.[25] This treatment was safe and resulted in some immunomodulating effect. Vaccination with clones of disease-inducing T cells obtained by antigen-specific expansion would be expected to be more effective than vaccination with cells obtained by non-specific expansion with a mitogen. However, in most human autoimmune diseases, information about the relevant autoantigen(s) is still too limited for the preparation of specific vaccines from disease-inducing T cells.

T cells recognize antigens as oligopeptides bound to MHC molecules via heterodimeric (αß or γδ) TCRs for antigen. Functional genes for the four different polymorphic TCR protein chains are specifically rearranged and expressed in T cells. They consist of a constant and a variable portion. The genetic information for the variable portion on each chain is generated by rearrangements of variable (V), diversity (D, only for ß and δ chains) and joining (J) gene segments. The variable gene segments together determine the antigen specificity of a T cell. A huge number of TCRs can be created by random combination of a relatively limited number of gene segments. Variable region diversity is created both by combinatorial assortment of V, D and J segments and by the loss or addition of nucleotides at their junctions. Studies in animal models of induced autoimmune diseases have strongly stimulated interest in TCR repertoires in human autoimmune diseases. In particular, the finding that T cells specific for MBP, which can be used to induce allergic encephalomyelitis in rats and mice, express a very limited repertoire of homologous Vα and Vß gene segments, has provoked much interest ('V-region disease hypothesis').[26,27] In animal models of induced autoimmune diseases a TCR analysis can be focused on T cells that are specific for the inducing antigen. Pathogenicity of the T cells analyzed can be demonstrated by their introduction into naïve hosts. In most human autoimmune diseases the nature of the relevant autoantigen is not known, so complete T cell infiltrates from sites of inflammation have to be analyzed. These are expected to consist of relevant pathogenic T cells and a background of lymphocytes that have been recruited by non-antigen-specific mechanisms. These 'bystander' cells will display a full range of TCRs, and their presence could obscure the overutilization of one or a few V gene segments by disease-inducing T cells. In fact, even in animal models in which T cells with specificity for the inducing antigen express conserved TCRs, an analysis of infiltrating cells at the site of inflammation does not necessarily reveal a V gene segment dominance.[28] In view of this it is not surprising that TCR repertoire studies in RA[25,29] and in other autoimmune diseases have yielded conflicting results. It is remarkable, however, that the analyses of synovial T cells in RA revealed dominating T cell clones in a polyclonal background in most cases.[29] Furthermore, clusters of patients may be recognized who share increased usage frequencies of certain V gene segments.[25,29,30] In MS, dominating TCR gene rearrangements have been identified in T cells specific for an immunodominating region of MBP.[31,32]

These findings raised the question whether immunotherapeutic approaches could be directed against TCR gene products of disease inducing T cells (Table 2). Immunization with TCR peptides was successfully used to influence autoimmune pathology by eliciting immune responses to determinants of the TCRs of disease-mediating T lymphocytes in experimental autoimmune encephalomyelitis (EAE).[33,34] In the same disease model, treatment with monoclonal antibodies specific for dominating TCR V gene segments was also successful.[26,35] Both types of TCR-targeted immunointervention have also been shown to be effective in animal models of other autoimmune diseases.[3] Even if TCR usage in human autoimmune diseases is less restricted than in animal models, treatment with cocktails of antibodies to a number of different V gene segments may be an option for a partially specific immunotherapy.

TARGETING THE PEPTIDE–MHC COMPLEX Susceptibility to various autoimmune diseases is linked to the expression of a given HLA haplotype. In animal models of autoimmune diseases, relationships between susceptibility and the expression of certain MHC gene products have been demonstrated, and treatment with antibodies specific for these gene products has been successfully used for experimental immunotherapy.[3] In principle, such treatment would down-regulate only those immune responses which are dependent on the targeted MHC molecule. Monoclonal antibodies specific for unique antigenic sites generated by binding of autoantigen- derived peptides to the MHC gene product[36] would be expected to have a more specific immunoregulatory activity. For specific immunotherapy of human autoimmune diseases with antibodies specific for HLA gene products, a severe restriction in the number of relevant HLA alleles would be a requirement (Table 2).

The T cell epitopes of autoantigens are in general very attractive potential targets for immunotherapy (Table 2), because T cell clones with specificity for certain peptide-MHC complexes could be specifically eliminated or made unresponsive by a treatment which is autoantigen-specific. The use of synthetic peptides to block interactions of autoantigens and MHC molecules or TCRs has been investigated in a variety of animal models of autoimmune diseases.[3,32] Tolerance to specific antigen determinants has been successfully induced by administration of self-peptides in non-immunogenic formulations, and with complexes of soluble MHC class II molecules loaded with such pathogenic peptides. Peptide analogues which bind to the disease-linked MHC antigen with high affinity, but lack T cell activating potential, were also effective in blocking disease-causing peptide-MHC antigen interactions. Furthermore, competitor peptides which are structurally unrelated to the disease-causing peptide have also been successfully used in animal models of autoimmune diseases, and attempts are currently being made to identify non-peptidic chemical entities with specific MHC blocking activity. The mechanisms by which TCR antagonists operate remain to be determined. Induction of tolerance by mechanisms other than MHC blockage may play a role in certain types of peptide-induced unresponsiveness.

Lack of knowledge about the relevant autoantigens in human autoimmune diseases has so far limited the clinical evaluation of these antigen-specific therapeutic approaches. However, there is evidence that the major autoimmune response in MS is directed to a region of myelin basic protein between residues 84 and 103, and it has been suggested that responses to other myelin antigens arise from intermolecular spreading of the immune response at the site of inflammation.[32] Plans exist to tolerize MS patients to the MBP- derived peptide p84-102 with an altered peptide ligand. Results obtained in EAE indicate that it may be possible to reverse the pathogenic autoimmune process by such a specific immunotherapeutic approach. It is hoped that by inducing tolerance to MBPp84-102, autoimmune responses to other epitopes of MBP or even other myelin antigens can also be downregulated.[32] In another related approach, a random copolymer of tyrosine, alanine, lysine and glutamate has been used in a pilot clinical trial in patients with relapsing-remitting MS.[37]

TOLERANCE INDUCTION BY ORAL ADMINISTRATION OF ANTIGEN It has long been known that a state of systemic hyporesponsiveness to an antigen can be induced by oral administration.[38] Depending on the dose administered, clonal anergy or active suppression mediated by the induction of regulatory cells in the gut-associated lymphoid system may be of primary importance. Oral tolerance has been successfully used to treat autoimmune phenomena in various animal models (Table 4). Human trials have been initiated in MS[39] with bovine myelin protein, in RA with type II chicken colla-

TABLE 4 Immunotherapy by oral administration of antigen

	Condition	Fed antigen
Animal models	Experimental allergic encephalomyelitis (EAE)	Myelin basic protein (MBP), proteolipid protein (PLP)
	Experimental arthritis	Type II collagen
	Experimental autoimmune uveoretinitis (EAU)	S-antigen, interphotoreceptor retinoid-binding protein (IRBP)
	Diabetes	Insulin, glutamate decarboxylase
	Myasthenia gravis	Acetylcholine receptor
	Thyroiditis	Thyroglobulin
Human autoimmune diseases	Multiple sclerosis (MS)	Bovine myelin
	Rheumatoid arthritis (RA)	Chicken type II collagen
	Uveoretinitis	Bovine S-antigen

gen[40] and in uveitis with bovine S-antigen.[41] In the initial phase I/II trials no apparent toxicities or exacerbations of disease were observed. Since the initial results were encouraging, larger double blind controlled studies have been initiated.[38] The identification of all antigen(s) involved in the progression of an autoimmune disease may not be a prerequisite for the induction of oral tolerance, because bystander suppression may be operative as long as regulatory cells specific for the fed antigen will locate to the site of autoimmune inflammation. Suppression may be most effective with homologous protein, which indicates a possible future requirement for large amounts of human recombinant proteins for oral immunotherapy.

Immunotherapy of uveitis

Experimental autoimmune uveoretinitis (EAU) can be induced by immunization of rodents or primates with purified retinal antigens, such as the retinal soluble antigen (S-antigen), rhodopsin or the interphotoreceptor retinoid-binding protein (IRBP). In these model systems, many strategies of partially specific and specific immunotherapy have been used with some success.[42] Adhesion molecule blockade with antibodies to ICAM-1 and LFA-1 protected mice against development of EAU after IRBP-immunization.[43] Targeting of activated T cells with an interleukin-2/diphtheria-toxin fusion protein, 7-10 days after immunization with S-antigen, suppressed EAU development in Lewis rats.[44] Inhibition of EAU was correlated with a deletion of lymph node cells capable of adoptively transferring EAU. Protective effects against EAU induction by transfer of S-antigen-specific T cells were also observed with IL-2 receptor-specific antibodies in combination with a subtherapeutic dose of cyclosporin.[45]

T cell vaccination by injection of IRBP-specific T cells had some protective effect against challenge with an uveitogenic dose of the same cells.[46] Because TCR Vß gene segment usage by uveitogenic T cells seems to be restricted in EAU,[47,48] attempts were made to use Vß8–derived peptides to vaccinate against T cells that express the implicated TCR. Results were variable, ranging from weakly protective to even stimulatory effects.[49] It is remarkable that with identical TCR peptides a far stronger suppression of experimental autoimmune pathology was observed in EAE.[34] Treatment of rats with MHC class II blocking antibodies resulted in a downregulation of S-antigen induced autoimmune uveoretinitis.[50] Peptides with high affinity for MHC class II molecules also downregulated EAU.[51] Oral administration of S-antigen specifically suppressed S-antigen induced EAU in Lewis rats.[52] Protection was also achieved by feeding a synthetic peptide containing the dominant pathogenic epitope, but not by feeding peptides containing only subdominant pathogenic epitopes.[52,53]

Oral administration of bovine S-antigen is also now being evaluated in clinical uveitis.[41] Cyclosporin has already been used successfully in treating severe sight-threatening disease and a number of additional medications such as FK506, rapamycin and mycophenilate mofetil are currently being tested.[42] However, at present, the most common mode of therapy of clinical uveitis is still treatment with systemic corticosteroids.[42] In the future, progress in the specific immunotherapy of other autoimmune diseases may also stimulate the development of new treatment modalities for uveitis.

Acknowledgement

We would like to thank Jennifer Jenkins for careful reading of this manuscript and useful discussions.

References

1 Filaci G, Zanetti M. Therapy of autoimmune diseases. In: Bona CA, Siminovitch KA, Zanetti M, Theofilopoulos AN, editors. The Molecular Pathology of Autoimmune Diseases. Chur: Harwood Academic Publishers, 1993:777-92.

2 Adorini L, Guéry J-C, Rodriguez-Tarduchy G, Trembleau S. Selective immunosuppression. TiPS 1993;14:178-82.

3 Gaur A, Fathman CG. Immunotherapeutic strategies directed at the trimolecular complex. Adv Immunol 1994;56:219-65.

4 Lehmann PV, Sercarz EE, Forsthuber T, Dayan CM, Gammon G. Determinant spreading and the dynamics of the autoimmune T-cell repertoire. Immunol Today 1993;14:203-8.

5 Pluschke G, Rüegg D, Hohlfeld R, Engel AG. Autoaggressive myocytotoxic T lymphocytes expressing an unusual γ/δ T cell receptor. J Exp Med 1992;76:1785-9.

6 Bach J-F. Immunosuppressive therapy of autoimmune diseases. TiPS 1993;14:213-16.

7 Veys EM, Mielants H, Vergruggen G. Thymic hormones and interferons in treatment of rheumatoid arthritis. Scand J Rheumatol 1988;5:297-304.

8 Fathman CG, Myers BD. Cyclosporin therapy for autoimmune disease. New Engl J Med 1992;326:1693-4.

9 Zanetti M. Idiotypy and the idiotype network in autoimmunity. In: Bona CA, Siminovitch KA, Zanetti M, Theofilopoulos AN, editors. The Molecular Pathology of Autoimmune Diseases. Chur: Harwood Academic Publishers, 1993:209-28.

10 Elliott MJ, Maini RN. New directions for biological therapy in rheumatoid arthritis. Int Arch Allergy Immunol 1994;104:112-25.

11 Waldmann H, Cobbold S. The use of monoclonal antibodies to achieve immunological tolerance. TiPS 1993;14:143-8.

12 Lockwood CM, Thiru S, Isaacs JD, Hale G, Waldmann H. Long-term remission of intractable systemic vasculitis with monoclonal antibody therapy. Lancet 1993;341:1620-1.

13 Van der Lubbe PA, Dijkmans BAC, Markusse HM, Nässander U, Breedveld FC. Lack of clinical effect of CD4 monoclonal antibody therapy in early rheumatoid arthritis: a placebo-controlled trial. Arthr Rheum 1994;37:S294.

14 Moreland L, Pratt P, Mayes M, Postlethwaite A, Weisman M, Schnitzer T, Lightfoot R, Calabrese L, Koopman, W. Minimal efficacy of a depleting chimeric anti-CD4 (cM-T412) antibody in treatment of patients with refractory rheumatoid arthritis receiving concomitant methotrexate. Arthr Rheum 1993;36:S39.

15 Olsen NJ, Crush JJ, Lipsky PE, St Clair EW, Cannon G, MacCune WJ, Strand V, Lorenz T. Multicenter trial of an anti-CD5 immunoconjugate in

rheumatoid arthritis (RA). Arthr Rheum 1994;37:S295.

16 Strom TB, Kelley VR, Murphy JR, Nichols J, Woodworth TG. Interleukin-2 receptor-directed therapies: Antibody- or cytokine-based targeting molecules. Annu Rev Med 1993;44:343-53.

17 Moreland L, Sewell KL, Sullivan WF, Shmerling RH, Parker KC, Swatz WG, Woodworth TG, Trentham DE, Koopman WJ. Double-blind placebo-controlled phase II trial of diphtheria-interleukin-2 fusion toxin (DAB486II-2) in patients with refractory rheumatoid arthritis (RA). Arthr Rheum 1993;36:S39.

18 Sharar SR, Winn RK, Harlan JM. The adhesion cascade and anti-adhesion therapy: an overview. Springer Semin Immunopathol 1995;16:359-78.

19 Maini RN, Elliott MJ, Brennan FM, Williams RO, Qiu Chu C, Paleolog E, Charles PJ, Taylor PC, Feldmann M. Monoclonal anti TNFα antibody as a probe of pathogenesis and therapy of rheumatoid disease. Immunol Rev 1995;144:195-223.

20 Rankin EC, Choy EHS, Kassimos D, Soowith M, Kingsley G, Isenberg DA, Panayi GS. A double-blind, placebo-controlled, ascending dose trial of the recombinant humanised anti TNFα antibody CDP571 in patients with rheumatoid arthritis (RA): A preliminary report. Arthr Rheum 1994;37:S295.

21 Moreland L, Margolies GR, Heck LW, Saway PA, Jacobs C, Beck C, Blosch C, Koopman WJ. Soluble tumour necrosis factor receptor (sTNFR): results of a phase I dose-escalation study in patients with rheumatoid arthritis. Arthr Rheum 1994;37:S295.

22 Cohen IR, Weinter H. T cell vaccination. Immunol Today 1988;11:332-5.

23 Zhang J, Medaer R, Stinissen P, Hafler D, Raus J. MHC-restricted clonotypic depletion of human myelin basic protein-reactive T cells by T cell vaccination. Science 1993;261:1451-2.

24 Medaer R, Stinissen P, Truyen L, Raus J, Zhang J. Depletion of myelin-basic-protein autoreactive T cells by T-cell vaccination: pilot trial in multiple sclerosis. Lancet 1995;346:807-8.

25 Breedveld FC, Struyk L, Van Laar JM, Miltenburg AMM, De Vries RRP, Van den Elsen P. Therapeutic regulation of T cells in rheumatoid arthritis. Immunol Rev 1995;144:5-16.

26 Heber-Katz E, Acha-Orbea H. The V-region disease hypothesis: evidence from autoimmune encephalomyelitis. Immunol Today 1989;10:164-9.

27 Heber-Katz E, Acha-Orbea H. Reply. Immunol Today 1993;14:380-2.

28 Bell RB, Lindsey JW, Sobel RA, Hodgkinson S, Steinman L. Diverse T cell receptor Vß gene usage in the central nervous system in experimental allergic encephalomyelitis. J Immunol 1993;150:4085-92.

29 Pluschke G, Krawinkel U, Hohlfeld R. T cell receptor repertoires in rheumatic diseases. In: Zierhut M, Thiel H-J, Saal J, editors. Immunology of the joint and the eye. Buren: Æolus Press, 1996: 189-201.

30 Grom AA, Thompson SD, Luyrink L, Passo M, Choi E, Glass DN. Dominant T-cell-receptor ß chain variable region $Vß14^+$ clones in juvenile rheumatoid arthritis. Proc Natl Acad Sci USA 1993;90:11104-8.

31 Oksenberg JR, Panzara MA, Begovich AB, Mitchell D, Erlich HA, Murray RS, Shimonkevitz R, Sherritt M, Rothbard J, Bernard CCA, Steinmann L. Selection for T-cell receptor Vß-Dß-Jß gene rearrangements with specificity

for a myelin basic protein peptide in brain lesions of MS. Nature 1993;362:68-70.
32 Steinman L, Waisman A, Altmann D. Major T-cell responses in MS. Molec Med Today 1995;1:79-83.
33 Howell MD, Winters ST, Olee T, Powell HC, Carlo DJ, Brostoff SW. Vaccination against experimental allergic encephalomyelitis with T cell receptor peptides. Science 1989;246:668-70.
34 Vandenbark AA, Hashim G, Offner H. Immunization with a synthetic T-cell receptor V-region peptide protects against experimental autoimmune encephalomyelitis. Nature 1989;341:541-4.
35 Zaller DM, Osman G, Kanagawa O, Hood L. Prevention and treatment of murine experimental allergic encephalomyelitis with T cell receptor Vß-specific antibodies. J Exp Med 1990;171:1943-55.
36 Aharoni R, Teitelbaum D, Aron R, Puri J. Immunomodulation of experimental allergic encephalomyelitis by antibodies to the antigen-Ia complex. Nature 1991;351(6322):147-50.
37 Bornstein MB, Miller A, Slagle S, Weitzman M, Crystal H, Drexler E, Keilson M, Merriam A, Wassertheil Smoller S, Spada V, et al. Clinical experience with COP-1 in multiple sclerosis. Neurology 1988;38(7 Suppl 2):66-9.
38 Weiner HL. Oral tolerance. Proc Natl Acad Sci USA 1994;91:10762-5.
39 Weiner HL, Mackin GA, Matsui M, Orav EJ, Khouri SJ, Dawson DM, Hafler DA. Double-blind pilot trial of oral tolerization with myelin antigens in multiple sclerosis. Science 1993;259:1321-4.
40 Trentham DE, Dynesius-Trentham RA, Orav EJ, Combitchi D, Lorenzo C, Sewell KL, Hafler DA, Weiner HL. Effects of oral administration of type II collagen on rheumatoid arthritis. Science 1993;261:1727-30.
41 Nussenblatt RB, De Smet MD, Weiner HL, Gery I. In: Wechsler B, Godeau P, editors. Proceedings 6th International Conference on Behcet's Disease. Amsterdam: Elsevier, 1993.
42 Caspi RR, Nussenblatt RB. The natural and therapeutic control of ocular autoimmunity: rodent and man. In: Autoimmunity: Physiology and Disease. Chichester: Wiley Liss Inc., 1994:377-405.
43 Whitcup SM, DeBarge LR, Caspi RR, Harning R, Nussenblatt RB, Chan CC. Monoclonal antibodies against ICAM-1 (CD54) and LFA-1 (CD11a/CD18) inhibit experimental autoimmune uveitis. Clin Immunol Immunopathol 1993;67:143-50.
44 Roberge FG, Lorberboum-Galski H, Le Hoang P, De Smet M, Chan CC, Fitzgerald D, Pastan I. Selective immunosuppression of activated T cells with the chimeric toxin IL-2-PE-40. Inhibition of experimental autoimmune uveoretinitis. J Immunol 1989;143:3498-3502.
45 Higuchi M, Diamantstein T, Osawa H, Caspi RR. Combined anti-interleukin-2 receptor and low-dose cyclosporine therapy in experimental autoimmune uveoretinitis. J Autoimmun 1991;4:113-24.
46 Beraud E., Kotake S, Caspi RR, Oddo SM, Chan CC, Gery I, Nussenblatt RB. Control of experimental autoimmune uveoretinitis by low dose T cell vaccination. Cell Immunol 1992;140:112-22.
47 Egwuagu CE, Chow C, Beraud E, Caspi RR, Mahdi RM, Brezin AP, Nussen-

blatt RB, Gery I. T cell receptor beta-chain usage in experimental autoimmune uveoretinitis. J Autoimmun 1991;4:315-24.
48 Gregerson DS, Fling SP, Merryman CF, Zhang XM, Li XB, Heber-Katz E. Conserved T cell receptor V gene usage by uveitogenic T cells. Clin Immunol Immunopathol 1991;58:154-61.
49 Kawano Y, Sasamoto Y, Kotake S, Thurau SR, Wiggert B, Gery I. Trials of vaccination against experimental autoimmune uveoretinitis with a T-cell receptor peptide. Curr Eye Res 1991;10:789-95.
50 Wetzig R, Hooks JJ, Percopo CM, Nussenblatt R, Chan CC, Detrick B. Anti-Ia antibody diminishes ocular inflammation in experimental autoimmune uveitis. Curr Eye Res 1988;7:809-18.
51 Kozhich A, Vistica P, Gery I, Berzofsky J, Wauben M, Eden WV. Inhibition of experimental autoimmune uveoretinitis (EAU) by synthetic peptides. Invest Ophthalmol Vis Sci 1992;33:1025.
52 Nussenblatt RB, Caspi RR, Mahdi R, Chan CC, Roberge F, Lider O, Weiner HL. Inhibition of S-antigen induced experimental autoimmune uveoretinitis by oral induction of tolerance with S-antigen. J Immunol 1990;144:1689-95.
53 Thurau SR, Chan CC, Suh E, Nussenblatt RB. Induction of oral tolerance to S-antigen induced experimental autoimmune uveitis by a uveitogenic 20mer peptide. J Autoimmun 1991;4:507-16.
54 Guéry J-C, Sette A, Leighton J, Dragomir A, Adorini L. Selective immunosuppression by administration of major histocompatibility complex (MHC) class I binding peptides. I. Evidence for in vivo MHC blockade preventing T cell activation. J Exp Med 1992;175:1345-52.
55 Steinman L, Rosenbaum JT, Sriram S, McDevitt HO. Proc Natl Acad Sci USA 1981;78:7111-4.
56 Alexander J, Snoke K, Ruppert J, Sidney J, Wall M, Southwood S, Oseroff C, Arrhenius T, Gaeta FC, Colon SM, et al. Functional consequences of engagement of the T cell receptor by low affinity ligands. J Immunol 1993;150:1-7.
57 De Magistris MT, Alexander J, Coggeshall M, Altman A, Gaeta FC, Grey HM, Sette A. Antigen analog-major histocompatibility complexes act as antagonists of the T cell receptor. Cell 1992;68:625-34.
58 Clayton JP, Gammon GM, Ando DG, Kono DH, Hood L, Sercarz EE. Peptide-specific prevention of experimental allergic encephalomyelitis. Neonatal tolerance induced to the dominant T cell determinant of myelin basic protein. J Exp Med 1989;169(5):1681-91.
59 Gaur A, Wiers B, Liu A, Rothbard J, Fathman CG. Amelioration of autoimmune encephalomyelitis by myelin basic protein synthetic peptide-induced anergy. Science 1992;258:1491-4.
60 Sharma SD, Nag B, Su XM, Green D, Spack E, Clark BR, Sriram S. Antigen-specific therapy of experimental allergic encephalomyelitis by soluble class II major histocompatibility complex-peptide complexes. Proc Natl Acad Sci USA 1991;88(24):11465-9.
61 Cohen IR. Regulation of autoimmune disease – physiological and therapeutic. Immunol Rev 1986;94:5-21.
62 Zhang ZY, Lee CS, Lider O, Weiner HL. Suppression of adjuvant arthritis in Lewis rats by oral administration of type II collagen. J Immunol 1990;145:

2489-93.
63 Brod SA, al-Sabbagh A, Sobel RA, Hafler DA, Weiner HL. Suppression of experimental autoimmune encephalomyelitis by oral administration of myelin antigens: IV. Suppression of chronic relapsing disease in the Lewis rat and strain 13 guinea pigs. Ann Neurol 1991:29:615-22.

List of authors

Judith Alferink
Div. Molecular Immunology
Tumor Immunology Program
Deutsches Krebsforschungszentrum
Im Neunenheimer Feld 280
D-69120 Heidelberg
Germany
Fax +49-6221 401629

Bernd Arnold
Div. Molecular Immunology
Tumor Immunology Program
Deutsches Krebsforschungszentrum
Im Neunenheimer Feld 280
D-69120 Heidelberg
Germany
Fax +49-6221 401629

Evelyne Cash
INSERM U.283
Hôpital Cochin
27, rue du Fg. Saint-Jacques
75674 Paris Cedex 14
France
Fax +33-1-43 261156

Jeannine Charreire
Director, INSERM U.283
Hôpital Cochin
27, rue du Fg. Saint-Jacques
75674 Paris Cedex 14
France
Fax +33-1-43 261156

Dr. Gerald Coakley
Dept. Molecular Immunogenetics
17th Floor, Guy's Tower
Guy's Hospital
London Bridge, London SE1 9RT
England, U.K.
Fax +49-171 407 6689
E-mail g.coakley@umds.ac.uk

Prof. Antonio Coutinho, M.D.
Unité d'Immunobiologie
CNRS URA 1961
Institut Pasteur
Paris, France
Fax +33-1-45 688921

Andrew Dick, M.D.
Dept. Ophthalmology
University of Aberdeen Medical School
Foresterhill
Aberdeen AB9 2ZD
Scotland, U.K.
Fax +44-1224 685158

Ulrich Feige, Ph.D.
Dept. Pharmacology
AMGEN Center
Mailstop 15-2-A
1840 DeHavilland Drive
Thousand Oaks, CA 91320-1789
U.S.A.
Fax +1-805-447 1939
E-mail ufeige@amgen.com

Prof. Dr. T.E.W. Feltkamp
Ophthalmic Research Institute (IOI)
P.O. Box 12141
1100 AC Amsterdam
The Netherlands
Fax +3120 6916521

Prof. John V. Forrester, M.D.
Dept. Ophthalmology
University of Aberdeen Medical School
Foresterhill
Aberdeen AB9 2ZD
Scotland, U.K.
Fax +44-1224 685158

Jill Gasser
Dept. Pharmacology
AMGEN Center
Mailstop 15-2-A
1840 DeHavilland Drive
Thousand Oaks, CA 91320-1789
U.S.A.
Fax +1-805-447 1939

M.A. Hall
Dept. Molecular Immunogenetics
17th Floor, Guy's Tower
Guy's Hospital
London Bridge, London SE1 9RT
England, U.K.

Prof. Dr. Günter J. Hämmerling
Div. Molecular Immunology
Tumor Immunology Program
Deutsches Krebsforschungszentrum
Im Neunenheimer Feld 280
D-69120 Heidelberg
Germany
Fax +49-6221 401629

Prof. Günther Jung, Ph.D.
Institute of Organic Chemistry
Eberhard-Karls-Universität
D-72076 Tübingen
Germany
Fax +49-7071 296925

Bernhard Kreutzer, M.D.
Universitäts-Augenklinik
Schleichstr. 12
D-72076 Tübingen
Germany
Fax +49-7071 293730

Andreas Kuhröber
Institute for Med. Microbiology
Dept. Bacteriology, University of Ulm
Albert-Einstein-Allee 11
D-89081 Ulm
Germany
Fax +49-731 5024612

Maria Kuppner, Ph.D.
Dept. Ophthalmology
University of Aberdeen Medical School
Foresterhill
Aberdeen AB9 2ZD
Scotland, U.K.
Fax +44-1224 685158

Dr. Jerry S. Lanchbury
Guy's St. Thomas Hospital
Medical & Dental School
University of London
London SE1 9RT
England, U.K.

Andreas Limmer
Div. Molecular Immunology
Tumor Immunology Program
Deutsches Krebsforschungszentrum
Im Neunenheimer Feld 280
D-69120 Heidelberg
Germany
Fax +49-6221 401629

Janet Liversidge, Ph.D.
Dept. Ophthalmology
University of Aberdeen Medical School
Foresterhill
Aberdeen AB9 2ZD
Scotland, U.K.
Fax +44-1224 685158

Lynne Lumsden
Dept. Ophthalmology
University of Aberdeen Medical School
Foresterhill
Aberdeen AB9 2ZD
Scotland, U.K.
Fax +44-1224 685158

Georg Malcherek, M.D.
Neurologische Klinik
Kliniken Schnarrenberg
Eberhard-Karls-Universität
Hoppe-Seyler-Str. 3
D-72076 Tübingen
Germany
Fax +49-7071 296507

Asst. Prof. Elisabeth Märker-Hermann, M.D.
I. Dept. Medicine
Johannes-Gutenberg-Universität
D-55101 Mainz
Germany
Fax +49-6131 176609

Inga Melchers, Ph.D.
Klin. Forschergruppe f. Rheumatologie
Klinikum der Albert-Ludwigs-Universität
Breisacher Str. 62 (Neurozentrum)
D-79106 Freiburg
Germany
Fax +49-761 270 5298
E-mail melchers@nzll.ukl.uni-freiburg.de

Asst. Prof. A. Melms, M.D.
Neurologische Klinik
Kliniken Schnarrenberg
Eberhard-Karls-Universität
Hoppe-Seyler-Str. 3
D-72076 Tübingen
Germany
Fax +49-7071 296507

Karine Mignon-Godefroy
INSERM U.283
Hôpital Cochin
27, rue du Fg. Saint-Jacques
75674 Paris Cedex 14
France
Fax +33-1-43 261156

Prof. Claudia Müller, M.D.
Dept. Internal Medicine
Eberhard-Karls-Universität
D-72076 Tübingen
Germany
Fax +49-7071 293671

Asst. Prof. Gerd Pluschke, Ph.D.
Schweizerisches Tropeninstitut
Socinstrasse 57
CH-4002 Basel
Switzerland
Fax +41-61 2718654

Asst. Prof. Michael Raizman, M.D.
New England Eye Center
Tufts University School of Medicine
750 Washington St.
Boston, MA 02111
U.S.A.
Fax +1-617-956 4866

Prof. Dr. Hans-Georg Rammensee
Dept. Immunology
Interfakult. Institut Zellbiologie
Eberhard-Karls-Universität
Auf der Morgenstelle 15
D-72076 Tübingen
Germany
Fax +44-7071 295653
E-mail rammensee@uni-tuebingen.de

Jörg Reimann, M.D.
Institute for Med. Microbiology
Dept. Bacteriology, University of Ulm
Albert-Einstein-Allee 11
D-89081 Ulm
Germany
Fax +49-731 5024612

Ortwin Rott, M.D.
(deceased, Sept. 1995)

Reinhold Schirmbeck
Institute for Med. Microbiology
Dept. Bacteriology, University of Ulm
Albert-Einstein-Allee 11
D-89081 Ulm
Germany
Fax +49-731 5024612

Alexej Schmidt
Neurologische Klinik
Kliniken Schnarrenberg
Eberhard-Karls-Universität
Hoppe-Seyler-Str. 3
D-72076 Tübingen
Germany
Fax +49-7071 296507

Rainer Stiemer, Ph.D.
Universitäts-Augenklinik
Schleichstr. 12
D-72076 Tübingen
Germany
Fax +49-7071 293730

Prof. J. Wayne Streilein, M.D.
Dept. Ophthalmology, Harvard Med. School
Schepens Eye Research Institute
20 Staniford St.
Boston, MA 02114
U.S.A.
Fax +1-617 720-1069

Lila Takorabet
INSERM U.283
Hôpital Cochin
27, rue du Fg. Saint-Jacques
75674 Paris Cedex 14
France
Fax +33-1-43 261156

Prof. H.-J. Thiel, M.D.
Universitäts-Augenklinik
Schleichstr. 12
D-72076 Tübingen
Germany
Fax +49-7071 293730

Robert Weissert
Neurologische Klinik
Kliniken Schnarrenberg
Eberhard-Karls-Universität
Hoppe-Seyler-Str. 3
D-72076 Tübingen
Germany
Fax +40-7071 296507

Wolfgang Wienhold
Neurologische Klinik
Kliniken Schnarrenberg
Eberhard-Karls-Universität
Hoppe-Seyler-Str. 3
D-72076 Tübingen
Germany
Fax +40-7071 296507

Asst. Prof. Manfred Zierhut, M.D.
Universitäts-Augenklinik
Schleichstr. 12
D-72076 Tübingen
Germany
Fax +49-7071 293730

Subject index